Advance praise for Nexus | NEW INTERSECTIONS IN INTERNET RESEARCH |

"Researching socio-technological systems is never a simple task. The complications, the social and technical relations, and their ever-changing nature makes this a challenge for any researcher. But in this volume, several individuals and teams are taking up the challenges of this nexus and bringing their research to fruition through interdisciplinary, and perhaps even transdisciplinary efforts. These authors' efforts all derive from the Oxford Internet Institute's summer doctoral programme, which is a programme that I am happy to be affiliated with as a participant in 2004 and an instructor in 2009. The topics the authors present derive from that program, but also go beyond it by creating new points of research analysis and revitalizing old ones with new research. Through these authors' efforts, this book reveals significant trends in Internet research and provides well-grounded insights into contemporary phenomena surrounding the Internet."

—*Jeremy Hunsinger, Virginia Tech*

"If this collection is a barometer of the future of Internet research in the next decade, then it's clear that we're in good hands. These collaborative contributions by young scholars from many countries are a tribute to the innovative programs of the Oxford Internet Institute and to the state of interdisciplinary Internet scholarship. In a time of rapid technological, social and economic change it's a relief to know that the kids are all right."

—*Larry Gross, Professor and Director, School of Communication,*
The Annenberg School for Communication & Journalism, University of Southern California

"The talented young scholars behind this collection break new ground in examining emerging collaborative uses of the Internet and the impact of such use across a range of settings. Well-written and thematically coherent, this volume takes us deeper into the social implications of ICTs, while at the same time ably demonstrating the multidisciplinary and transnational character of the developing field of Internet studies."

—*Peter Dahlgren, Professor Emeritus, Media and Communication Studies,*
Lund University, Sweden

"*Nexus* brings together a range of studies and research by an international group of younger scholars that investigate and report on developments of digital technologies, focusing on intersections of new forms of collaboration with social, technical and political practices and the transformative effects of ICTs on all aspects of life in the twenty-first century. This is a collection of twelve essays written by digital natives who have not only grown up with the Internet but chosen it as a field of advanced study. The result is a stunning, wide-ranging, thought-provoking collection that focuses on topics including digital agency in relation to women, the future of news, media literacy, cyberinfrastructure, networked political activism, cybercrime and the green economy. This is a nicely crafted and well-edited volume that raises the stakes in Internet research."

—*Michael A. Peters, Professor of Educational Policy, University of Illinois at Urbana-Champaign*

"In this volume, the Oxford Institute and QUT at Brisbane have brought together a virtual 'dream team' of doctoral students, many of whom are already highly accomplished researchers. The main value of this volume is its strong contribution to empirical research on the subject of Internet research along with a healthy dose of reality-check on the promises and warnings surrounding cyber-discourse. This is a stellar collection with talented editors and deserves a close look by Internet researchers, activists and advocates."

—*Michel Bauwens, Founder, P2P Foundation; Associate Professor, ARC Centre of Excellence for Creative Industries and Innovation*

"Internet studies are structurally under threat of being outpaced by their object of study. Information and communication technologies, their adoption and the resulting uses are evolving at a rate which is simply incompatible with the normal pace of academic research. The Oxford Internet Institute and the talented Ph.D. students gathered within the yearly summer doctoral programme are not only incarnating the leading-edge research in the area, they are inventing new creative dynamics capable of keeping up with the pace and magnitude of changes related to the Internet. This book materializes both the means and the substance of a much-needed fresh approach to Internet research."

—*François Heinderyckx, Professor and Director, Department of Information and Communication Sciences, Université Libre de Bruxelles*

Nexus

This books is part of the Peter Lang Education list.
Every volume is peer reviewed and meets
the highest quality standards for content and production.

PETER LANG
New York • Washington, D.C./Baltimore • Bern
Frankfurt • Berlin • Brussels • Vienna • Oxford

Nexus

New Intersections in Internet Research

EDITED BY Daniel Araya,
Yana Breindl, Tessa J. Houghton

PETER LANG
New York • Washington, D.C./Baltimore • Bern
Frankfurt • Berlin • Brussels • Vienna • Oxford

Library of Congress Cataloging-in-Publication Data

Araya, Daniel.
Nexus: new intersections in Internet research /
Daniel Araya, Yana Breindl, Tessa J. Houghton.
p. cm.
Includes bibliographical references.
1. Internet—Social aspects. 2. Cyberspace—Social aspects.
3. Digital media. I. Breindl, Yana. II. Houghton, Tessa J. III. Title.
HM851.A73   303.48'33—dc22   2010046071
ISBN 978-1-4331-0970-6

Bibliographic information published by **Die Deutsche Nationalbibliothek**.
Die Deutsche Nationalbibliothek lists this publication in the "Deutsche
Nationalbibliografie"; detailed bibliographic data is available
on the Internet at http://dnb.d-nb.de/.

Cover concept by Tessa J. Houghton;
design by Daniel Araya

The paper in this book meets the guidelines for permanence and durability
of the Committee on Production Guidelines for Book Longevity
of the Council of Library Resources.

© 2011 Peter Lang Publishing, Inc., New York
29 Broadway, 18th floor, New York, NY 10006
www.peterlang.com

All rights reserved.
Reprint or reproduction, even partially, in all forms such as microfilm,
xerography, microfiche, microcard, and offset strictly prohibited.

Printed in the United States of America

Contents

Foreword
RALPH SCHROEDER   VII

Acknowledgments   xi

Introduction: Collective Intelligence
DANIEL ARAYA   1

Section One: Sociocultural Intersections

1. Women and Technology: "Five Acts of Digital Agency"
   ANITZA GENEVE & CARLA GANITO   13

2. Snap, Post, Share: Understanding the Online Social Life
   of Personal Photography
   ERIC COOK & CRISTINA GARDUÑO FREEMAN   35

3. Geo-linguistic Analysis of the World Wide Web: The Use of
   Cartograms and Network Analysis to Understand Linguistic
   Development in Wikipedia
   THOMAS PETZOLD & HANTENG LIAO   55

4. Mapping the Future of News in a Digital World: US and Australian
   Perspectives
   LUCY MORIESON & NIKKI USHER   77

Section Two: Technosocial Intersections

5. Media Literacy in the Facebook Age: Designing Online and Face to Face Learning Environments
   ANDRÉS MONROY-HERNÁNDEZ, MICHAEL DEZUANNI, & KAI KUIKKANIEMI   95

6. eHealth: Bridging the Divide between Current Performance and Legitimate Expectations in Health Care Delivery
   LUCA CAMERINI & YUJUNG NAM   117

7. Fielding Networked Marketing: Technology and Authenticity in the Monetization of Malaysian Blogs
   JULIAN HOPKINS & NEAL THOMAS   139

8. Cyberinfrastructure Inside Out: Definition and Influences Shaping Its Emergence, Development, and Implementation in the Early 21st Century
   KERK KEE, LUCY CRADDUCK, BRIDGET BLODGETT & RAMI OLWAN   157

Section Three: Political Intersections

9. Leetocracy: Networked Political Activism or the Continuation of Elitism in Competitive Democracy
   YANA BREINDL & NILS GUSTAFSSON   193

10. Haxorz: Alternate Perspectives on the "Computer Underground"   213

    The People's Republic of Hacktivism: A Public Sphere Theoretical Interpretation of Online Independence Movements and the People's Republic of China
    TESSA J. HOUGHTON   215

    Cybercrime: A New Challenge for Legislation and International Negotiation
    YAO-CHUNG CHANG   230

11. ICTs and the Green Economy: US and Chinese Policy in the 21st Century
    DANIEL ARAYA, JIN SHANG, & JINGFANG LIU   239

    Afterword
    JEAN BURGESS & MARCUS FOTH   255

    Contributors   261

Foreword

RALPH SCHROEDER

It is very exciting to see the doctoral students from the Oxford Internet Institute's Summer Doctoral Programme (OII SDP) producing such a compelling and timely volume on *new intersections in Internet research*. To give some background on our summer programme at Oxford, OII brings together doctoral students from around the world to study with leading academics in the field of Internet Studies, and provides an academic framework in which to share and discuss students' current research. OII SDP has now been going since 2003, alternating each year between Oxford and another country. So far we have paired up with the Chinese Academy of Social Sciences in Beijing (2005) and the Berkman Center at Harvard University (2007). In July 2009, the SDP was held in Brisbane and hosted by the Creative Industries Faculty at Queensland University of Technology (QUT). This volume is the happy outcome of that intensive two-week period, followed by months of online collaboration.

The theme for 2009 was 'Creativity, Innovation and the Internet' and QUT was the ideal place to explore this important theme. Like OII, the researchers at QUT are highly multidisciplinary, with a focus on new media and creative expression. The summer school in Brisbane was noteworthy, among other things, for featuring an extremely international programme of lecturers. They included Jack Linchuan Qiu from the Chinese University of Hong Kong, Nancy Baym from University of Kansas, and Jeremy Hunsinger from Virgina Tech. From QUT we had talks by Stuart Cunningham, John Hartley, Axel Bruns, Brian Fitzgerald, Jo

Tacchi and many others. Marcus Foth and Jean Burgess did a fantastic job of both organizing a lively programme of talks and activities and leading discussions on their own pioneering research. And it was a great pleasure for me to be joined in Brisbane by OII'ers Bernie Hogan and Rebecca Eynon.

Year after year, students have said that the most valuable aspect of the SDP is that they are able to discuss their dissertation topics with students and faculty that are—like them—working on Internet-related research. In their home universities and home disciplines, there are often few (if any) other students or faculty working on Internet topics. At the SDP, doctoral researchers can benefit from being immersed in the company of other like-minded (or at least like-focused) researchers. At the same time, this two-week immersion in a diversity of topics and approaches is a great way to expand one's approach to a given topic (and one's vocabulary—among other things, I learned of the word 'heteroclite'—which certainly applies to this diverse collection of essays).

This volume covers a rich array of topics and themes: from new phenomena such as Flickr, Wikipedia and other forms of user generated content, to the application of recent theories like gender and agency, actor-network-theory, and webometrics, to a variety of current domains such as hacktivism, the use of cyberinfrastructure in research, green policies, e-Health and online journalism. The volume also reflects a range of countries being studied and students from around the world. Together, these young researchers exemplify the strengths of multidisciplinarity. Here I pick just out one essay which is close to my own current research area—cyberinfrastructure or e-Research: In this chapter, there is a combination of sociology of science and technology with law and public policy that yields a fresh look at this topic, unlike most research in this area which usually either has a sociology of science or a policy audience in mind.

What is especially pleasing to see in these contributions is that the pursuit of PhD research, which is typically a highly personal and often a lonely pursuit, has here been turned into a collective and collaborative enterprise. The resulting essays have been enriched, not just by multiple authorship, but also by combining perspectives and topics. To use the terminology of the volume, these are *new intersections*. Anyone studying the social aspects of the Internet will know that it's a hard job keeping up with the thing—it's a fast-moving target. This collection demonstrates the terrific variety and dynamism of Internet research, and the benefits of forging ahead into uncharted territories.

Bringing together the work of the SDP students ensures that the network of Internet research keeps expanding. If you, reader, go to the annual meeting of the Association of Internet Researchers (AoIR), you will find that it is partly an SDP 'class reunion'. I look forward to this book being a focal point at this event and in Internet studies generally. It remains for me to thank colleagues, especially at QUT, and students for taking part in a lively and memorable summer programme, and to thank the students for giving me the opportunity to write this short foreword. I wish them every success in their careers as Internet researchers. If this collection of essays is anything to go by, they are sure to shape the field of social research on the Internet for the foreseeable future.

Ralph Schroeder
Oxford Internet Institute
(Oxford, February 2010)

Acknowledgments

Nexus could not have been written without the help of many people and institutions. We would like to express our gratitude to all of them.

Our endeavor to create a collaborative volume on the current state and future of Internet research would not have been conceivable without the Oxford Internet Institute (OII), who run the Summer Doctoral Programme (SDP) that brought us all together in 2009. As such, we owe the OII a debt of gratitude for providing this opportunity—especially Bill Dutton, Victoria Nash, Ralph Schroeder, Rebecca Eynon, Bernie Hogan, and Laura Taylor.

Jean Burgess, Marcus Foth, and Jaz Choi did an amazing job of hosting the SDP09 at the lovely Queensland University of Technology (Kelvin Grove) campus, and of ensuring we had a great Brisbane experience. They kept us in line, and, in the words of Marcus, "focused," despite such distractions as the Etherpad backchannel and Rhonda. They also challenged us to create a legacy project—this volume is (one of) the outcomes of that challenge.

Special thanks go to the spontaneous outcry of Ralph Schroeder—"This should be a book!"—during one of the student collaboration sessions. His suggestion and subsequent support, as well as that of Rebecca Eynon, Jean Burgess, and Marcus Foth, have shaped the successful completion of this volume.

We also wish to thank the tutors who attended the SDP09 for sharing their research and ideas with us, and stretching our brains till they hurt (in the best way possible!):

Nancy Baym, Stuart Cunningham, Gerard Goggin, John Hartley, Jack Linchuan Qiu, Stephanie Hemelryk Donald, Axel Bruns, Jo Tacchi, Brian Fitzgerald, Jean Burgess, Marcus Foth, Ralph Schroeder, Rob Ackland, Matthew Allen, Wendy Seltzer, Rebecca Eynon, Jason Wilson, Kate Crawford, Larissa Hjorth, Bernie Hogan, and Jeremy Hunsinger.

Each of us is very grateful to those who suggested, encouraged, supported, and helped fund our participation in the SDP09: our supervisors, advisors, colleagues, universities and grant institutions. Thanks for making this possible! The details of these supportive agents are as follows:

Supervisors and advisors:

Adrian McCullagh, Andrea Tapia, Anne Fitzgerald, Brian Fitzgerald, Cathy Greenfield, Darin Barney, Donald Matheson, Francois Heinderyckx, Isabel Gil, Jan Teorell, Kate Sweetapple, Larry Gross, Linda-Jean Kenix, Magnus Jerneck, Margaret McLaughlin, Martha Pollack, Michael A. Peters, Michel Bauwens, Mitchell Resnick, Muhammed Musa, Naomi Stead, Patti Riley, Peter Grabosky, Peter J. Schulz, Stephanie Teasley, Steve Jackson, and Yeoh Seng Guan.

Universities and grant institutions:

Catholic University of Portugal (Research Centre for Communication and Culture); Intel Foundation, Iomega; Lund University (Department of Political Science); Massachusetts Institute of Technology (Media Lab); University of Illinois at Urbana-Champaign (College of Education); McGill University (Art History and Communication Studies); Microsoft; Monash University (Sunway Campus, School of Arts & Social Sciences); Nokia; Penn State University (College of Information, Science and Technology); Queensland University of Technology (Faculty of Law & Creative Industries and the Faculty of Science and Technology); RMIT University (Media and Communication); Siamon Foundation; Universite Libre de Bruxelles; United States National Science Foundation Grants; University of Canterbury (Media and Communication); University of Lugano (Institute of Communication and Health); University of Michigan (School of Communication & Rackham Graduate Programme); University of Southern California; University of the Sunshine Coast (Faculty of Business); University of Sydney, Univeristy of Technology, Sydney, and the Wahlgren Foundation.

We (individually) would like to thank each other for our amazing collaborative effort since July 2009. Starting with finding a (or several) partner(s); writing an abstract; creating the *Nexus* proposal; and finally writing the chapters across continents, oceans and time zones! Meeting each other was the best part of SDP09!

Special thanks to Daniel Araya for magically finding us an editor—Peter Lang—in an extraordinarily short amount of time; and to Peter Lang for publishing *Nexus*.

Introduction

Collective Intelligence

DANIEL ARAYA

Taken together, the global reach of digital technologies and the rising influence of networked collaboration constitute a new mode of information production that is reshaping industrial societies. Beyond the command-and-control systems characteristic of industrialization, information and communication technologies (ICTs) have become fundamental to a network age (Castells, 1996). Underlying this socioeconomic restructuring is the critical importance of digital networks as platforms for creative collaboration.

New tools inevitably engender changes in the way people interact, communicate, and collaborate (Wenger, 1998). However, it is increasingly clear that ICTs are now leveraging a unique democratic shift in a wide array of technological, political, and social spaces. This edited collection interrogates the current ways mass collaboration intersects with sociocultural, technosocial, and political changes in the context of new social practices. Intertwined with these new practices are questions about the impact of ICTs on established institutions and modes of production.

Much like the invention of the printing press, the Internet is radically transforming the most basic elements of modern civilization. The success of mass collaboration in a multitude of contexts poses a challenge, not only to the dominant economic paradigm, but also to a broad range of received social science thinking. There is mounting evidence that new forms of social networking bind people together in highly creative social and economic relationships. People are no longer passive participants in their own economic and cultural production, but are becoming active agents in the production of lived social and political environments.

THE RISE OF COLLECTIVE INTELLIGENCE

For many, the Internet represents a sociotechnological platform on top of which the knowledge, resources, and computing power of millions of people are coming together into a massive collective force (Tapscott & Williams, 2006). Just as new systems of meaning-making emerged with the printing press, the Internet is generating new cultural forms, and reorganizing the basic mechanisms of cultural power. As Yochai Benkler (2006) has pointed out, the distributed nature of ICTs is giving rise to a democratic shift in "peer production" that is opening cultural life to new modes of creativity and innovation. This trend is not only obvious in the context of software and information production (most notably in the free and open-source software movement) but also physical production in the context of user-led innovation (Von Hippel, 2005).

The growing opportunities for amateur producers to work together in scaled collaboration are changing institutions as widespread as journalism, education, government, healthcare, communications, and entertainment. Energized through peer-to-peer collaboration, the Internet is enabling new forms of decentralized and amorphous self-organization that is impacting a wide range of institutions and practices (Leadbeater, 2000). As these effects permeate contemporary society and intersect with deep structural changes in the global economy, we are observing a significant restructuring of mass industrial society. Researchers are now challenged to develop new models for describing and understanding these changes. What theories and tools do we have today to help us explain this cultural shift?

One of the more ambitious attempts to interpret these trends is Pierre Lévy's (1997) early work on the subject of "collective intelligence." As Lévy suggests, collective intelligence underlies a new paradigm that is emerging in various fields of research simultaneously,

> Far from being exclusive, the expression "collective intelligence" relates to an extensive body of knowledge and thoughts concerned with several objects that have been diversely labeled: distributed cognition, distributed knowledge systems, global brain, super-brain, global mind, group mind, ecology of mind, hive mind, learning organization, connected intelligence, networked intelligence, augmented intelligence, hyper-cortex, symbiotic man, etc. Notwithstanding their diversity, these several rich philosophical and scientific contemporary trends have one feature in common: they describe human communities, organizations and cultures exhibiting "mind-like" properties...(p. 1)

For Lévy, the World Wide Web represents the emergence of a semantic commons that is gradually enabling the whole of humanity to house and manage its cultural heritage. As sociocultural theorists suggest, human cognition is indistinguishable from ongoing sociocultural practices because it is anchored to social tools and artifacts (Lave, 1988; Lave & Wenger, 1991; Varela, Thomson, & Rosch, 1991). For these researchers, whole systems of artifacts (words and numbers) form the basic foundations for shared cultural cognition. Just as new tools of labor facilitate new social structures, new tools of thinking facilitate new cognitive structures. This tool-mediated understanding of human cognition is even more obvious today. With the emergence of worldwide ICT networks, systems of cultural cognition are becoming increasingly global (Robertson, 1992).

THE GLOBAL MAP

There are now an estimated 1.4 billion Internet users in the world today, with growth in developing countries expanding at five times the rate of developed countries (developing countries now account for more than half of the world total of Internet users) (UNCTAD, 2009, p. 11). China is now the world's single largest broadband market and has the largest number of Internet users (298 million), followed by the United States (191 million) and Japan (88 million) (ibid.). Wide gaps in ICT infrastructure (especially in broadband networks) remain, however. While the Internet is undoubtedly expanding in reach and density, access remains fragmented and uneven. Gaps in broadband connectivity between high-income and low-income countries, for example, are only widening. Average broadband penetration is more than eight times higher in developed than in developing countries.

Many people now see technology as a key to negotiating problems in the twenty-first century. While technology may solve some problems, however, it can also magnify others. Today the vast majority of the world's population remains disconnected from ICTs. Access to ICTs is largely divided along economic lines. While more than half the population in the developed world is now online, only 15 to 17 percent of people in developing and transition economies are online (UNCTAD, p. 11). There is reason to suggest that this is not a permanent trend, however. The spread of mobile telephony, for example has been explosive. There are more than four billion mobile phone subscribers in the world today:

> On average, there are now 60 subscriptions per 100 people, and in many developed, developing and transition economies penetration exceeds 100. Reflecting explosive growth, the penetration level in developing countries is now eight times higher than what it was in 2000. Almost every second person in developing countries is thought to have a mobile phone and

fewer than a dozen developing nations have a mobile penetration of less than ten. Between 2003 and 2008, the most dynamic economies in terms of increased mobile penetration were outside the developed world. (UNCTAD, 2009, p. 11)

Part of this is explained by the high rates of economic growth in many emerging economies. One can only assume that the impact of emerging economies in the twenty-first century will be considerable. By 2050 the combined economies of Brazil, Russia, India, and China could eclipse the combined economies of the current richest countries. China already has the largest foreign exchange reserves and it has overtaken the United States and Germany to become the world's largest exporter. In 2009, China became the second largest economy in the world and it is predicted to become the world's largest (followed by India) by the middle of this century.

COMPLEXITY AND INTERNET RESEARCH

ICTs are proliferating at an incredible rate. Consider, for example, the incredible growth of the World Wide Web. The Web represents an emergent phenomenon that is highly resistant to linear models of inquiry. Researchers today are beginning to develop new models for describing and understanding these complex dynamics. One important resource for modeling this is found in the science of complex systems. While traditional methods of research often assume that systems under study are characterized by some kind of order and stability, complex systems are often unpredictable. At no point do complex systems come to a natural equilibrium or stasis.[1]

Unpredictability and change are critical to understanding complex systems because transformation is often a consequence of the feedback arising from the shifting relationship of agents within the system. Complex systems are "heterarchical" systems in which the parts of a network (individuals, groups, etc) are greater than their sum. This includes the behavior of those parts and the emergent behavior of the system as a whole (Laszlo, 1996). Theorists in the social sciences using arguments based in complexity suggest that linear methods of analysis miss embedded elements such as contingency and multicausality. Studies anchored in a positivist methodology, for example, often overlook the situated environment in which systems unfold and develop (Barnes, Matka, & Sullivan, 2003).[2]

Complexity science offers one of the few tools we have for understanding the interactivity and flux that make up the social dimensions of Internet-related phenomenon. "*Think of the spontaneous organization* of half a million ants or termites, which allows them to construct complex hills and nests" (Ferguson, 2010, p. 24). Nothing commands the Internet; instead, the interaction of a multitude of individual agents combine to form a higher level of collective intelligence. While peer producers may be structurally dependent on technological systems, for example, they are also critical agents in the ongoing construction of those systems (Giddens, 1984).

The value of complexity science to Internet research is that it offers a dynamic middle ground between the reductionism of positivism and the ambiguity of poststructuralism (Bhaskar, 1997; Haynes, 2007). In the context of this volume, complexity thinking is useful for understanding the emergent forms of collective intelligence that underlie a large degree of Internet-related phenomena. These emergent intelligences are critical to understanding the reason Internet research is interesting to so many researchers across the disciplinary spectrum.

Nexus: New Intersections in Internet Research

To better contextualize the chapters found in this volume, we must first recognize the tremendous diversity of disciplines that come under the umbrella of Internet research. The term "Internet Studies" is used in a flexible way by many different disciplines and for different purposes. In many ways, this disciplinary fragmentation reflects the origins of the Internet itself. The Internet consists of networks of interdependent nodes that conform to simple rules of behavior, but together exhibit emergent, system-wide behavior.

While simultaneously drawing and building from other research streams (computer-supported cooperative work, human–computer interaction, cyberculture, digital culture, new media studies, etc.), Internet research continues to evolve into what can only be called a meta-field of study. In this regard, this volume does not focus simply on a single key area but instead, seeks to comprehend the relationships, intersections, and interdependencies between multiple technology-mediated domains. The contributors represent a wide spectrum of disciplines but share a strong premise that ICTs are introducing discrete and coherent changes in varied environments.

The various approaches in this volume challenge readers to think outside disciplinary and methodological boundaries while at the same time recognizing an underlying unity that links the research. While some authors expose the social, political, and economic inequities that are often amplified by technology, others examine the widespread structural changes to institutions, practices, and policymaking. While studies of the Internet are now widespread across academic disciplines, there is a growing collaboration between investigators. This volume is one such example in what is hoped will be an ongoing multidisciplinary and transdisciplinary trend.

Organization of this Book

Section One: Sociocultural Intersections

Section One explores the *Sociocultural Intersections* of Internet research. New tools engender changes in the way people and communities interact, communicate, and construct their identities. In this section, we explore the ways in which ICTs enable transformations in existing sociocultural practices.

In Chapter 1, Geneve and Ganito provide insight into two empirical studies exploring women and technology. They explore women's participation through a lens of agency, where they argue that identities are influenced by both enablers and constraints to conform (to) or transform such influences. They provide an account of women's experiences with technology within a specific social context, through the conceptual categories of the Five Acts of Agency and emerging theory of Digital Agency.

In Chapter 2, Garduño and Cook argue that social networks such Flickr function as public repositories of personal photography that challenge and democratize the authority of traditional institutions of memory, like libraries and museums. Vernacular photography on these systems increases the visibility of individual's self-representation. In contrast to institutional collections, members on Flickr self-select the subject matter in their photographic contributions, the methods of representation, and the modes of participation.

In Chapter 3, Petzold and Liao look at linguistic aspects of the Internet and consider methods to measure this kind of activity on a large scale by using tools that can help generate this information. Preliminary analysis shows, for example, a relatively low overlap between Asian and European language versions of Wikipedia. Thus, they suggest, the concept and measurement of linguistic connectedness between Wikipedia nodes (different language versions) requires further explanations for the field of Internet studies.

In Chapter 4, Morieson and Usher consider the possible range of outcomes that can occur when the institution of news and the communicative possibilities of the Internet intersect. They explore the important nexus of cultural production and new communication technology, by mapping the current state of the media in the United States and Australia. Their aim is to provide a broad overview of the issues facing news as an institution, with particular reference to the threats and opportunities afforded by the development of the Internet.

Section Two: Technosocial Intersections

Section Two examines *Technosocial Intersections*. ICTs have become integral parts of a wide array of sectors. Networked technologies are being used for educational purposes or in order to improve health care delivery. They challenge the traditional media system, and certain phenomena such as personal blogging give birth to new commercial practices. This chapter discusses the wider repercussions ICT use has had on these domains.

In Chapter 5, Monroy-Hernández, Dezuanni, and Kuikkaniemi explore learning technologies, focusing on the ways young people participate in online and face-to-face spaces and how this can inform the design of social and creative learning technologies. In this context they discuss three specific social environments where young people create interactive media and gain new media literacy skills.

In Chapter 6, Camerini and Nam explore the topic of Internet technology in the health communication domain, generally referred to as eHealth. They lay out the implications, positive and negative, of a proliferating adoption of innovative technologies in bridging the divide between current performance and legitimate expectations in health care delivery.

In Chapter 7, Hopkins and Thomas locate and analyze a particular nexus in Malaysian blogging practices: the monetization of blogs, in particular the emergence of the "Lifestyle" blog. Drawing upon long-term ethnographic research, they consider the dynamics of this process by alternately using actor-network theory (ANT) and Bourdieuian field theory to trace relevant material-semiotic traffic between actors. They conclude by highlighting how social-symbolic capital exchange underwrites "authenticity" in Malaysian blogging, coming to focus also on how server logs quantify this capital circulation.

In Chapter 8, Kee, Cradduck, Blodgett, and Olwan provide a theoretically generative definition of cyberinfrastructure (CI) by drawing from existing definitions and literature in social sciences, law, and policy studies. They propose two models of domestic and international influencers on CI emergence, development, and implementation in the early 21st century. Based on its historical emergence and computational power, they argue that cyberinfrastructure is built on, and yet distinct from, the current notion of the Internet. The authors seek to answer two research questions: firstly, what is cyberinfrastructure? And secondly, what national and international influencers shape its emergence, development, and implementation (in e-science) in the early 21st century? Additionally, consideration will be given to the implications of the proposed definition and models, and future directions on CI research in Internet studies will be suggested.

Section Three: Political Intersections

Section Three examines political intersections. ICTs engender changes in the way polities construct themselves. In this section, we explore research that examines the ways in which ICTs enable transformations in existing political and policy practices.

In Chapter 9, Breindl and Gustafsson explore the influence of the Internet on traditional decision-making. While these forms of networked political organisations are usually perceived as less hierarchical than traditional mobilising groups, they point out the fact that successful forms of networked digital activism can be heavily dependent on technical and networking skills. Rather than functioning as the base of more egalitarian politics, the growing importance of networked political activism aided by digital media may on the contrary create new elites.

In Chapter 10, Houghton and Chang examine alternate perspectives on hacking, hacktivism, and cybercrime. Houghton assesses the case of hacktivism concerning territorial disputes centering on the People's Republic of China, and contends that hacktivism is a democratic activity in that it constitutes the emergence of multiple "neo-Habermasian" counterpublic spheres. In contrast, Chang explores hacking through a cybercrime lens, assessing difficulties with legislation and institutional self-protection against cybercrime, and making suggestions for overcoming these challenges.

In Chapter 11, Araya, Shang, and Liu consider U.S. and Chinese policy in the context of green innovation and a green economy. Many countries today are aggressively pursuing green innovation strategies because of the potential social and economic benefits associated with harnessing green technologies. Looking critically at economic policies in the U.S. and China, the authors consider current strategies for advancing green innovation. Much as ICTs have underwritten globalization and reshaped industrial societies, they argue that ICTs are foundational to a green economy.

Summary

There is an emerging consensus that Internet research necessarily transcends single disciplines. Information and Communication Technologies are transforming the way we live, work, and play, and as research on the Internet converges with other disciplines, new discourses are forming regarding technology's growing significance. New tools inevitably engender changes in the way people interact, communicate and collaborate, and it is increasingly clear that the Internet is providing new spaces for research that require new models of interpretation. Just as new systems of meaning-making emerged with the printing press, the Internet is generating new cultural forms, and introducing new forms of cultural power. While there is still no common definition of the term *Internet Studies*, the authors in this volume have endeavored to contribute to and advance this growing field.

Notes

1. As Eoyang and Berkas (1998) observe, complex systems have more in common with permanent whitewater or unshackled action than predictability. The dynamic nature of complex systems means that researchers cannot isolate the underlying factors that determine outcomes. Since complex systems are open systems, boundaries must often be defined arbitrarily. Factors outside a system's boundaries may have as much influence on the behavior of a complex system as the dependent and independent variables within it.
2. Similarly, research linked to a managerial or policy focus often overestimates the capacity of human agency to steer complex systems (including the Internet).

References

Barnes, M., Matka, E., & Sullivan, H. (2003). Evidence, understanding and complexity: Evaluation in non-linear systems. *Evaluation. 9*(3) 265–284. London: Sage.
Benkler, Y. (2006). *The wealth of networks: How social production transforms markets and freedom.* New Haven, CT: Yale University Press.
Bhaskar, R. (1997). *A realist theory of science.* New York: Verso.
Castells, M. (1996). *The rise of the networked society.* Oxford: Blackwell.
Eoyang, G. & Berkas, T. (1998). Evaluation in a complex adaptive system. In Lissak & Gunz (Eds.), *Managing complexity in organisations.* Westport, CT: Quorum Books.
Ferguson, N. (2010). Complexity and collapse: Empires on the edge of chaos. *Foreign Affairs.* 89 (2), pp. 18-32.
Giddens, A. (1984). *The constitution of society: Outline of the theory of structuration.* Cambridge, UK: Polity
Haynes, P. (2007). *Complexity theory and evaluation in public management: A qualitative approach.* Health and Public Policy Research Centre, University of Brighton.
Laszlo, E. (1996). *The systems view of the world: A holistic vision for our time.* Cresskill, NJ: Hampton.
Lave, J. (1988). *Cognition in practice.* Cambridge, UK: Cambridge University Press.
Lave J. & Wenger E. (1991). *Situated learning: Legitimate peripheral participation.* New York: Cambridge University Press.
Leadbeater, C. (2000). *Living on thin air: The new economy.* London: Penguin.
Lévy, P. (1997). *Collective intelligence: Mankind's emerging world in cyberspace.* New York: Plenium.
Robertson, R. (1992). *Globalization: Social theory and global culture.* London: Sage.
Tapscott, D. (1996). *The digital economy: Promise and peril in the age of networked intelligence.* New York: McGraw-Hill.
UNCTAD (2009). *The* information economy report 2009*: Trends and outlook in turbulent times.* New York: United Nations.
United Nations Conference On Trade And Development (2009). Information economy report 2009: Trends and outlook in turbulent times. New York: United Nations.
Varela, F., Thomson, E., & Rosch, E. (1991). *The embodied mind: Cognitive science and human experience.* Cambridge, MA: The MIT Press.
Von Hippel, E. (2005). *Democratizing innovation.* Cambridge, MA: The MIT Press.

Section One

Sociocultural Intersections

CHAPTER ONE

Women and Technology

"Five Acts of Digital Agency"

ANITZA GENEVE & CARLA GANITO

Thompson (1984) suggests that "few questions remain as refractory to cogent analysis as the question of how, and in precisely the ways, the action of individual agents is related to the structural features of the society of which they are a part" (p. 148). This chapter provides a brief overview of how insights from two different but complementary empirical studies assist us to move closer to answering this question. Specifically, the research problem considers the inequitable participation by women in certain facets of contemporary life related to digital technology, tools, and artifacts. Such inequitable participation has existed for some time; that it continues in a time of profound technological and sociological change is a concern.

This chapter argues that a human agency perspective provides a theoretical approach to understand women's participation. Agency is viewed as Giddens (1984) suggests, as an individual's "capacity to act" or "act otherwise" in the face of certain conditions. An agency perspective explores the problem of women's participation without reverting to essentialist approaches to gender—focusing on the agent moves us away from the binary of male and female. Women are not helpless victims of a patriarchal structure, but rather constrained or enabled agents with the potential for change.

This chapter articulates a conceptual framework, the *Five Acts of Agency*, which draws on an agency perspective to understand women's actions, decisions, and participation. The framework emerges from Geneve's study of women's participation in the Australian Digital Content Industry (DCI) and is one of two conceptual tools that underpin an emerging theory of *Digital Agency*.[1] The *Five Acts of Agency* framework is used to explore the findings of the first case study, contextualized in the DCI. In a second case study, we explore the transferability of the *Five Acts of Agency* framework to a different context by applying it to Ganito's Portuguese study exploring women's uptake and use of mobile phones. Both empirical studies offer rich descriptive insights to complement the theoretical perspective.

THE AUSTRALIAN DIGITAL CONTENT INDUSTRY (DCI)

The DCI produces digital content such as websites, computer and console games, and mobile phone content. It is recognised as an important emerging industry within the Australian economy, an "innovation frontier" following Cunningham, Cutler, Ryan, Hearn, and Keane (2003), and one where there is an identified skills shortage in certain areas (DCITA, 2005; AIMIA, 2005). There is, however, an under-representation of women in interactive content creators role in Australia (Geneve, Nelson & Christie, 2009). Similarly, in the United States, Deuze, Martin, and Allen (2007) identify that "male workers almost completely dominate the core content creation roles (such as design, programming, and visual arts)" (p. 346). There is, however, limited research of this phenomenon, particularly of women's experiences in the DCI context. There are notable exceptions including, Gill (2002) and Roan and Whitehouse (2007). Consequently, the case study considers the body of literature related to women's participation in the related Information Communication and Technology industries (ICT).

Although the ICT industries and DCI share similarities, for example the occupational role of computer programmer exists in both, there are also differences. For example, previous studies have suggested that occupational roles in Information Technology (IT) are perceived (particularly by young people) as being geeky, boring, and involve working alone (Beekhuyzen & Clayton, 2004). DCI roles differ in that they demand passionate workers (Gill, 2007; Roan & Whitehouse, 2007) who are the "cool, creative, and egalitarian" (Gill, 2002) professionals of computer-related careers. The key similarity and concern between the industries is that women are under-represented in both.

Why Agency?

Previous research presents a litany of influences, including masculinization of the industry, a lack of role models, and stereotypes surrounding gendered identities. Geneve et al. (2009) identified several common themes surrounding these influences, including:

- Individual: behaviors, personality traits (e.g., confidence), skills, and perceptions;
- Social: cultural, historical (e.g., power relationships), media, family; and
- Environmental: resource access (e.g., equipment), industry characteristics (e.g., long hours).

In considering the "individual" influences, an agency perspective moves us away from the male/female binary or hardwired biological differences, often presented in the previous literature as a way of explaining differences in participation. In considering the "social" and "environmental" influences, an agency perspective recognises that the environment may well present deterrents. Ramsey and McCorduck (2005) note that for women in IT, "circumstances almost seem designed to wedge them from the work they love" (p. 7). Importantly, an agency perspective also recognises a women's capability to understand, challenge, and transform such influences. Therefore, agency offers a way to understand the circumstances or environment (contextualized historically by technological and social change) and an emergent socially situated interaction, where individuals exercise agency to meet their capabilities.

Furthermore, that previous studies have been positioned from different paradigmatic stances (Trauth & Howcroft, 2006) may contribute to what Quesenberry (2006) considers to be a wide range of fragmented and difficult to holistically understand findings. Theories of agency can bridge the constructivist, functionalist, and essentialist polemics and lead to a unified understanding of influences. Indeed, Giddens (1984) presents Structuration Theory as a bridge between ontological approaches, from structuralism (such as Parsons) to extreme individualism (favored by postmodernists).

Case Study Methodology

The exploratory case study, contextualised in the Australian DCI, involved interviewing 18 female interactive content creators in the multimedia and games production sectors over the 2007–2009 period. Their occupations involve the use of technology, such as computers and software, in the production of digital cultural products such as interactive content (websites and computer games). Examples of their occupational roles include artificial intelligence programmer, game designer, and web interaction designer. Data collection and analysis followed the procedures expected of a case study (see Eisenhardt, 1989; Stake, 1995; Walsham, 1995; Yin, 1994). The hermeneutic mode of analysis encouraged an understanding of conditions or circumstances from the female DCI professional's perspective, acknowledging her active role in understanding influences. As Layder's (1998) Adaptive Theory approach underpins the research process, existing theories "scaffold" (Walsham, 1995, p. 76) the analysis of empirical data and development of an emerging theory of *Digital Agency*.

Case study Outcome—The Emerging Theory of Digital Agency

Positioned from a critical realist stance, the emerging theory encourages a focus on interaction between the environment and the person across multiple levels of a stratified reality. This includes the agent's action, embedded in social practices that may constrain and enable the transformative capacity of such action. The approach proffered here acknowledges that the conditions of agency (both the circumstances and the agent's identity) will influence an agent's experience. The analytical process requires exploration of both the agent and the structures, in a way that avoids the criticism of conflation that Archer (1995) has levelled at Giddens (1984).

Although structures or structured practices may present similar circumstances for several agents, the response from individuals may vary. Such an approach resonates with Trauth's *Individual Differences Theory of Gender and IT* (Trauth et al., 2004, 2005). Following Giddens' ontology of potentials, the theory of *Digital Agency* acknowledges that every agent has a similar potential. However, the types of experiences and possibilities an agent may experience will be filtered to varying extents due to their identity. Identity here refers primarily to their social identity, often recognized by society through norms and stereotypes, rather than a self-identity. As Connell (2002) suggests, "gender is, above all, a matter of the social relations within which individuals and groups act" (p. 9).

The Sphere of Influence

Two conceptual tools emerge from the case study and provide a way to explore women's participation or agency. The first, the *Sphere of Influence* (Geneve et al., 2009), is what Giddens and Turner (1988) refer to as a "sensitizing analytical framework." It encourages the description of both the environment (where and when) and the agent's identity (who) so as to sensitize the researcher to the emergent interactions between these. It draws on extant theory, as a complementary analytical tool, to further explore these interactions. The theories include Giddens' (1984) Structuration Theory (to explore the processes between society and agent) and Bandura's (1997; 1999) Social Cognitive Theory (to explore the cognitive mechanisms of the agent).

The Five Acts of Agency

If the *Sphere of Influence* illuminates the conditions of the agent's experience, the *Five*

Acts of Agency brings to the foreground the individual's acts of agency. It asks the question, "How does an agent act (become aware of, conform, or transform these conditions) in response to constraints or supports?" Although Giddens' meta-theory proposes there are "rules" or "generalisable procedures applied in the enactment/reproduction of social life" (Giddens, 1984, p. 21), the empirical findings may provide insight into how the individuals transformative capability or "power to intervene" may "make a difference" (Giddens & Turner, 1988, p. 284). Figure 1 presents the elements of an "act of agency" as including:

~ conditions (the dimensions and properties of the environment);
~ an agent (understood as their individual capacity but also acknowledging the social norms surrounding a filtered identity e.g., gender);
~ interaction (between the environment and the actions by the agent);
~ reflectivity (the agents reflection on their actions or inactions, the conditions and their identity);
~ consequences (of the actions).

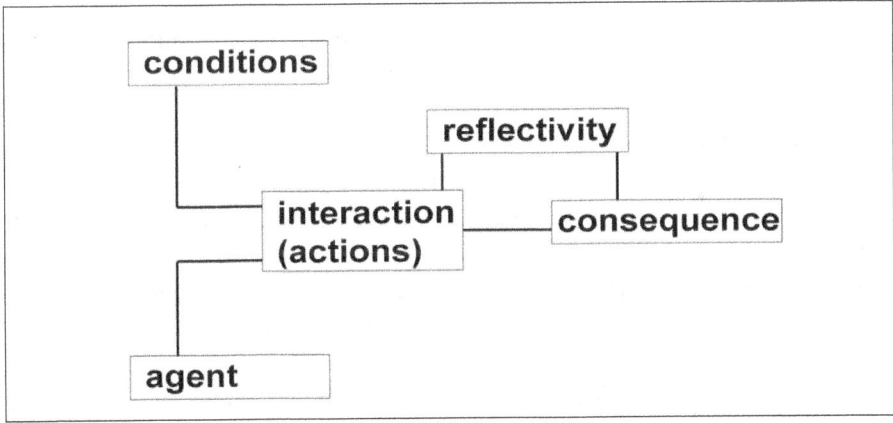

Figure 1. The elements of an 'Act of Agency'

The *Five Acts of Agency* is a heuristic typology of agentic "acts" (see Table 1). An agent may exercise their capability in five ways, through being: enabled, connected, collaborative, creative, and human (see Table 1 for sub-categories). Each type of act considers the conditions women face and the consequences of their actions (at times exercised as a choice of inaction). These acts are gendered only in the sense that these conditions are most likely to influence the social identity labeled woman. The next section utilizes the *Five Acts of Agency* framework to provide an insight into the influences on participation that woman working in the Australian Digital Content Industry have experienced.

Table 1: The Five Acts of Agency (and their sub-categories)

Acts of Agency				
Enabled agent Core tools and resources such as technology, people, and information.	Connected agent The beliefs and motivations that attract the agent, including self-efficacy.	Collaborative agent The social relationships that foster involvement.	Creative agent Problem solving or the power to transform.	Human agent Human only traits of emotions and ethics.
• Technology • Information • People	• Imagining • Doing	• Belonging • Sharing	• Problem solving • Creating/ transforming	• Emotive • Ethical • Being

ACTS OF AGENCY IN THE AUSTRALIAN DIGITAL CONTENT INDUSTRY (DCI)

The Enabled Agent

The category of enabled (technology, information, people) agent explores influences such as access to technological *resources*, a key aspect of the "digital divide" (see Castells, 1996; Kvasny & Trauth, 2002). It also encompasses resources such as *information* and *people*. For the female DCI professional, access to *enabling technology* was a prevalent influence in her childhood. Often other social agents, frequently fathers and brothers, introduced her to her first computer. This highlights a historical influence, as at the time these women were children technology was less available than it is currently. As adults in the industry, they have more ready access to the "allocative" (Giddens, 1984) resources of their occupation.

Participants who had made an explicit decision to pursue pathways to work in the DCI found initial entry one of the most difficult barriers to overcome, noting that social relationships and connections were a positive influence in gaining access. An example of "authorative" (Giddens, 1984) resources occurred in the form of the *people* who facilitated these women's access into the industry or as Bandura (2001) may describe them, proxy agents. The women reported that these connections or "informal networks" (Gill, 2002) could occur in unlikely places, for example, "having friends who are involved in the industry is a good way to start" as "unless you're really well networked it's quite difficult to get a leg in, in this city" (m5[2]).

Although presented here as distinct analytical categories, the "acts of agency" are relational. In everyday life there are no clear boundaries surrounding them. For example, to have access to people who could serve as proxy agents or mentors might also facilitate access to information and resources. However, having access to information in itself may not enable the agent to form a sense of connection to the industry career pathway.

The Connected Agent

The category of connected (imagining, doing) agent explores the influences that motivate an agent. It is not as much a social "connection" to others but a connection by the agent to the possibility of participating in the industry. This occurs through their identification with the domain. For example, suitable role models foster an agent visualizing or *imagining* herself as participating within an industry context.

> "You have to find a way to become involved and there are barriers to that because it's not something highly visible [...] I think there was a barrier there in my head. You know programming is sort of seen as you know a boy's thing . . ." (m5)

Agents may be motivated through vicarious observation of others obtaining rewards, such as monetary or the prestige of being "artistic, young, and 'cool'" (Gill, 2002, p. 28). Yet there is a paucity of occupational role models for young women in the emerging DCI. As one participant noted

> "there isn't an image of a hot female programmer out there that people want to emulate or aspire to" (g10).

The connected category also identifies that once an agent has access to resources, such as a computer, it is the act of gaining skills through *doing* that fosters further connection. A participant recounted the sense of self-reward when learning how to program from her father's computer magazines at a young age; she recalled it was like learning a recipe. The case study identifies that for the female DCI professional feeling connected hinges on their "self-efficacy" (Bandura, 1997) towards skills involving technology. Bandura (1997) suggests that the vicarious observation of role models, in particular same sex role models, is also an important aspect of an agent's self-efficacy. The women noted that social norms had led them to believe that computing skills are the domain of "boys." This may explain their sense of surprise when they discovered how easy it was to pick up and cultivate these skills:

> "If you got shown how to set up the programs and that, it's really not that hard." (g5)

Therefore, the connected agent develops a sense of self-efficacy. As various participants expressed, receiving positive feedback from other people may further enhance self-efficacy:

> "Management says good things and that was an enormous confidence booster, so yeah I kinda thought this is not so bad." (m5)

"Just the support and congratulations, I don't know, I just thought that was like wow, whether it was true or not I don't care but they came out and said done good." (m2)

Gürer and Camp (2002), in presenting the "shrinking pipeline" metaphor, suggest women are a minority in the ICT workplace due to a gradual decline of participation rates within contexts such as school and workplace. Similarly, Geneve et al., (2009) have identified that for women in the DCI, influences manifested over a lifespan, including childhood, educational, and workplace contexts. Relationships with technology begin from an early age and continue over their career. One participant noted that she taught herself "basic programming when I was about six" (g2).

The Collaborative Agent

The category of the collaborative (belonging, sharing) agent exposes influences such as social norms and practices, particularly those surrounding identity stereotypes, which manifest in social settings. Women working in the DCI often challenge the stereotypical norms regarding women and technology. In doing so, they transform not only their colleagues and societies assumptions of gender but also their own. Participants in the study identified themselves as being "kinda like one of the boys" (g2) and different to other women, with "other" women defined as being "girly girls" and nontechnical.

Such norms influence the sense of *belonging* a female DCI professional feels. Norms may manifest through something as everyday as the clothes one wears. One participant noted that the DCI industry allowed workers to wear whatever they wanted, which meant that you could "be comfortable all day and you can express yourself." She continued, "I can't imagine working somewhere where you're expected to wear a skirt" (g3). Giddens (1993) notes similar gender identification through clothing in a school setting, stating "regulations that compel girls to wear dresses or skirts in school form one of the most obvious ways in which gender typing occurs" (p. 43).

Sanctioning by one's peers in the workplace can create a sense of *belonging*. One participant recalled the impact on her self-confidence when recommended by a male social acquaintance for her current position:

"For me that was sort of a bit of an achievement because I actually felt confident probably for the first time in my whole career that I could actually do this work."(m5)

Being a minority in the workplace may adversely affect a sense of *belonging*. In one games company, female DCI professionals instigated "ladies lunches" in response to not only the low number of women employed in the organisation but also to the distribution of women in the work environment. The agent driven initiative was seen as important by the participants, as the company's rapid growth had lead it to become "so big all the girls had sort of been scattered around" and that they might not "even pass each other in the corridor," which led them to feel that they were in a "minority" (g1). These lunch events provided a regular opportunity to informally network with other women in the company. Another participant noted she did not think it was a problem that she was the only female programmer in her team. She suggested that this was because of the nature of the company she worked with, as they were a "very close together company." It emerged that she could overcome any sense of feeling isolated due to her gender by choosing to mix with other females (mostly from the web design and HTML coding team) over tea and lunch breaks. She noted that it might be more obvious she was the only "female sitting in a big room (…) if it's in a bigger company" particularly if "departments were isolated" (m3).

The collaborative act also entails the *sharing* of resources, expertise, and experience. The DCI industry is perceived as a close-knit community whose members share knowledge and resources such as software and code quite freely, where "everyone works off each other's back" (m5). Mentoring, as either mentor or mentee, is an example of such sharing. A participant noted how sharing "skills and knowledge" (m5) lead to a collective sense of pride towards the products they developed.

The Creative Agent

The category of creative (problem solving, transforming) agent acknowledges that "we create society at the same time as we are created by it" (Giddens, 1984, p. 14). The games and multimedia sectors may be considered as one of the "creative industries" (Cunningham et al., 2003). Roan and Whitehouse (2007) identify that both advanced programming skills and creative talent are essential aspects of the DCI worker. Consequently, creativity is a highly valued capability of individuals in the creative industries. This was also the case for the female DCI professional. Having the capacity to be creative greatly motivated their participation in the industry. The term creative is, however, also a term that is difficult to define and is beyond the scope of this discussion here. For the participants, creativity appeared to involve *problem solving*, the challenge of finding solutions to technical or client-based problems which in turn provided intrinsic reward.

> "There's such a variety with multimedia there's always something different to do in it, because you'll always have to learn new things."(m7)

> "And once you start learning and getting results and you actually make a program you can get a reward to say wow this is fun, I want to make it more complex." (g5)

Creativity also involves the *transforming* of what may be "everyday" to something different. For interactive content creation, this involves the creation of things such as games characters, a storyline, a design layout etc.

> "The whole creative of what something's going to look like, all of the early stages of the artwork because I love art and it's wonderful to see a character to go from say a green frog and it ends up being a purple dragon and just the changes." (g1)

A majority of participants identified themselves as always being "a creative person" including those from an IT background. Participants cited a lack of scope for creativity in their previous roles for leaving other industries, such as publishing.

The Human Agent

The category of human (emotive, ethical, being) agent is perhaps the hardest to explore in the research, due to its existentialist nature. This category prompts the researcher to consider: what brings meaning to the agent's lives once their needs are met in a Maslow like manner? Similar to previous studies in the DCI (Roan & Whitehouse, 2007; Gill, 2007) and IT context (Griffiths, Keogh, Moore, Tattersall, & Richardson, 2005), the women in the case study stated how "passionate" they were in regards to their occupations. Archer (2000) provides a place for such *emotions* in agency. The case study identified that emotions fuelled participation, yet also at times, could lead to employers exploiting their passionate employees.

> "It's a cycle in that you know everyone is passionate about work to put in these hours, so then it's expected, and so then it becomes the norm and then everyone else has to put in. It's a labour of love and that's the same sort of people you get to work ridiculous hours." (g12)

Although the above quote suggests long hours are a norm in the industry, several participants challenged the norm by actively choosing not to work longer hours than what they believed to be reasonable. This provides an empirical illustration of agents' exercising their power to, as Giddens (1984) describes, "act otherwise" (p.14). It also illustrates that an act involves not only actions but also reflection on the consequence of those actions. The women who choose not to work long hours felt sanctioned by their peers for not fitting in. This reinforces that "acts of agency" are interrelated; when agents addressed their emotive human needs, they weakened their sense of belonging as a collaborative agent.

In summary, this first case study explored women's participation as workers in the Australian DCI. Drawing on agency, as a theoretical approach, helps explain the rich descriptive insights from the empirical data. The *Five Acts of Agency* illustrates how women act as enabled, collective, collaborative, creative, and human agents when working in the DCI. It also highlights that these acts occur within their private lives and over a lifespan, for example, acts in their childhood influenced their participation later in life. Ganito's case study on how women use mobile phones in Portugal serves as a further example of how the framework of the *Five Acts of Agency* can be applied to women and technology studies.

CASE STUDY—MOBILE ACTS: THE GENDERED USES OF THE MOBILE PHONE IN PORTUGAL

This second case study explores mobile phone usage in Portugal. Like the presence of women working in the DCI industry, women's uptake and use of mobile phones challenges traditional stereotypes of the relation of women towards technology. Simultaneously, we find that although the use of mobile technology is a potential site for a similar change there are still constraints in the transformation of those traditional stereotypes, making visible all the contradictions in the social shaping of technology. Although the study's philosophical underpinning is critical feminism, drawing on the emerging theory of *Digital Agency* offers a new perspective in understanding these constraints and enablers. The application of the *Five Acts of Agency* framework to mobile phone usage in Portugal highlights both examples of women's transformation and emancipation and of the persistence of traditional gender stereotypes. This reinforces the fact that usage parity does not mean usage equality.

Case Study Methodology

In this section of the chapter, we present the results of the first phase of the case study, which is a mixed method study (Tashakkori & Teddlie, 1998). The aim of using mixed methods was to discover paradoxes and contradictions that theory underlines as inherent to the gender–technology articulation. The first phase was mainly quantitative and consisted of the analysis of two national surveys: Mobile Portugal (2007) and E-Generation (2007). The first studied the usage of mobile phones by the Portuguese population and the second was centered on media and technology use by children and teenagers between the age of 9 and 18 years. This phase will be followed by a qualitative development of the research through focus group and in-depth interviews. Applying the *Digital Agency* theory to the quantitative analysis tests how the theory articulates with different data, in a different context, and in the scope of a mixed method case study.

The Portuguese Context

Portugal has a very high penetration rate of mobile phones, one of the highest in Europe, and a high usage rate, especially among those born after 1967. The latest survey of mobile phone users in Portugal shows that there is little difference between men and women in the uptake of mobile phones: 57.7% women and 42.3% men (Cardoso, Gomes, Espanha, & Araújo, 2007). According to the same survey, the group of nonusers is mostly compounded of older people, female, with low-education levels and generally unemployed. This is a common trend around the world (Castells, Fernandez-Ardevol, Qiu, & Sey, 2007).

Hans Geser (2004) argues that the mobile phone "levels differences between boys and girls," contrary to other technologies that tend to accentuate them. Although there has been a rapid and seemingly egalitarian adoption of mobile phones relative to other technologies, differences in experience do remain. These may give us interesting insights about the mobile phone as a "technology of gender" (Lauretis, 2004). Teresa de Lauretis proposes gender as a product of various social technologies, among which media (such as television, mobile phones, radio, and newspapers) open up possibilities for transformation. Lauretis' concept of "technology of gender" is rooted in Foucault's "technology of sex," defined as "a set of techniques for maximizing life." Lauretis proposes that the construction of gender and its effects manifest at the "local" level of resistances, in subjectivity and self-representation (p. 18) that "technologies of gender" are concerned with:

> "Not only how the representation of gender is constructed by the given technology, but also how it becomes absorbed subjectively by each individual whom that technology addresses" (Ibid., p. 223).

Similarly, Donna Haraway (1991) defends the need to go beyond the critique of representation and to incorporate the female subject in its multiplicity and subjectivity. From these viewpoints, there is no room for the former feminist view of media as only a transmitter of distorted images of women and of audiences as passive receivers of those images. Rather we may propose that the "actor" becomes an "agent" who has the capacity to "make a difference" (Giddens, 1984, p. 14) to her own life and those of others. Here is where the *Acts of Agency* theory can provide an innovative framework. We will use it to analyze the gendered use of mobile phones in Portugal.

Acts of Agency in the Use of Mobile Phones in Portugal

Women are commonly patronized in their use of technology but device and content producers are increasingly finding that traditional stereotypes are no longer holding their place. The *Acts of Agency* theory helps to shed light into the contradictions that companies and researchers are finding in women's every day use of technology.

The enabled agent

Under the category of enabled agent we find that young girls are *technology*, enabled primarily because parents are active prescribers of technology for them. A higher percentage of girls[3], in contrast to boys, receive their first mobile phone from their parents (Cardoso, Espanha, & Gomes, 2007, p. 174). Girls also obtain them at a younger age: 75.6% of girls receive their first mobile phone at age thirteen or less compared to 69.4% of boys. This is deeply rooted in safety concerns but it also means that girls have the opportunity to interact with mobile phones from a young age and to consider it a natural part of their media ecosystem. The fact that parents seem to support the phone expenses of girls more (Cardoso, Gomes et al., 2007, p. 168) suggests that young women are "enabled," as they draw on the resources other people can provide.

The connected agent

Although women seem to be keen users of mobile phones, they show some resistance as connected agents. Access to the mobile phone technology is fostered by the mobile phone becoming a massified technology (ibid.). Indeed, mobile phones are cheaply and widely available and therefore there is no transgression for a young girl to *imagine* herself using a mobile phone. The trend of young girls adopting the mobile phone is reinforced by increased marketing initiatives towards them, with major mobile brands releasing phones targeted to young girls using brands such as "Hello Kitty" and "Pucca" as triggers for their interest. There are no similar products for boys.

When considering how women may connect through the action of *doing*, that is gaining skill mastery, we note differences in the usage of the phone. Boys use the mobile phone to play games (31.95% of boys against 21.3% of girls) (Cardoso, Espanha et al., 2007, p. 170) in contrast to girls who tend to send text messages to friends, family members, or acquaintances about no specific matter when they are waiting for someone or are alone. Men seem to make more diverse use of the available functionalities of their mobile phones (alarm clock, calculator, e-mail, MMS, games, applications) (Cardoso, Gomes et al., 2007, p. 52). The same is true for services; men are more prone to try new functionalities (30.2% of men against 22.8% of women).

More boys than girls have cameras on their mobile phones: 59.6% of boys and 44% of girls (Cardoso, Espanha et al., 2007, p. 204), and boys take and send more photos (ibid., pp. 205–206). Nevertheless, Lee (2005) shows how young South Korean women appropriate their camera phone for cultural production, despite the prevalence of advertising that shows men snapping pictures of women. In this context, women are using mobile phones to perform new meanings:

> "These women are not the mere owners of camera phones, but performers who create various cultural meanings. They develop a more intimate relationship with technology, challenge the convention of gaze, give meaning to what is taken, and circulate their own expressions" (ibid., p. 12).

Such a use was not identified in the Portuguese case study; however, the Portuguese context does not provide the same gender constraints as Korea and thus transgressive practices are expected to be subtler. They may emerge with other technologies such as the usage of portable PCs or digital cameras. Girls and women use mobile phones more in comparison with other technologies because they are personal devices, free of the constraints of communal usage or criticism for unskilled or inappropriate usage.

The collaborating agent

If as *connected agents* women lag behind, they stand out as collaborating agents. Mobile phone use may foster a women's sense of *belonging*. This *belonging* mostly manifests in relation to women's traditional roles as social organisers of the family. Women tend to speak more to family members: 74.1% of women against 64% of men (Cardoso, Gomes et al., 2007, p. 19).

A sense of *belonging* involves the agent being aware of social norms and the sanctions for transgressing these. Therefore, the uptake of the mobile phone may be partly due to the fact that women are not required to challenge or transform social norms surrounding gender. Rather, as recent studies have presented, the mobile phone is another site for men and women to perform their traditional gendered identity, manifesting as "activity and technological appropriation for men" and "dependency and domesticity for women" (Lemish & Cohen, 2005). Cardoso, Gomes et al. (2007) note similar traditional gender roles: men stress instrumental phone use and women use them as a medium for personal and emotional exchange. The studies show that women use mobile phones for longer conversations about personal and emotional purposes, while men use mobile phones more often, for shorter calls for professional and utilitarian reasons (Cardoso, Gomes et al., 2007; Geser, 2004; Ling, 2004), matching the stereotypes. The Portuguese survey concluded that men make more calls that are deemed professional and women see the mobile phone as an instrument for managing their personal life.

The creative agent

Following the same constraints found in the category of connected agents, women also seem to find some barriers as creative agents. The penetration of more sophisticated devices amongst women is still very low and thus their ability to produce mobile content is limited by access to devices with the appropriate capabilities. Women may express their creativity through the personalization of their mobile phone and mobile phone content. But personalization is more common amongst boys: 72.2% against 54.5% of girls (Cardoso, Espanha et al., 2007, p. 208) and men: 55% compared to 44% of women (Cardoso, Gomes et al., 2007, p. 208). Personalization by men includes choice of a screensaver or ringtone. The Portuguese survey findings suggest women are not interested in personalization. Nevertheless, the only personalization that was taken into account were pictures, screensavers, and ringtones that demand a higher technical proficiency and possibly a greater demand for *problem solving*. Other practices of personalization (such as covers and dangle cords) were not taken into account, and it may be the case that Portuguese women would prefer these personalization features that do not require sophisticated devices.

In contexts where female culture is more constrained, more creative practices can be found. In Japan, young girls use the mobile phone to challenge masculine culture in public places. These girls *transform* the constraints of their gendered identity through their creative acts:

> "Through their tactics of play, colorful dress, and mobile phone usage, they fight the hegemony of older men on subways, and in doing so produce space for themselves" (Steenson, 2006, p.8).

As women intensify their usage of technological artifacts, "we can see a transformation of women's interests stereotypes" (Skog, 2002, p. 268). Yet, in the meantime, constraints keep moving beyond stereotypical gendered identity construction. A more expansive expression of identity is constrained by the low access to more sophisticated devices. Furthermore, the device itself is marked by traditional gender "scripts." For example, identity can be expressed through color. Here the dichotomous nature of mobile phones comes up again. It is argued that turning them pink to appeal to women is a "gender script." The connection between pink and femininity has become second nature (Peril, 2002), a social norm to which we conform to avoid punishment and being shamed. When women choose pink mobile phones to highlight their presence or embody the nature of their positions as a form of subversion, they are also "articulating within strictly defined boundaries" (Puwar, 2004, p. 151). They are expressing their right to mobile technology but are stating their gender through stereotypes, because being too subversive would mean not being accepted. There is pressure on women to highlight their difference from men through exaggerated forms of femininity, as a mask to protect them from retaliation or to hide their intentions or their real power (Riviere, 1929; Puwar, 2004). The other option is to mimic the hegemonic culture of male dominated black, grey, and silver technological aesthetics. It becomes a trap for women, a risk in transgressing their gendered identity.

The human agent

For the human agent, mobile phone usage elicits an *emotive* response in the human, and these emotions enable or constrain the use of these technologies. The mobile phone is regarded as an affective technology (Plant, 2001; Lasen, 2004), an object of mediation, demonstration, and communication of feelings and emotions. This emotional connection translates into personalization, and as discussed earlier, this personalization fosters a sense of belonging, albeit often to a traditional gender identity. As noted earlier, mobile phones can provide a sense of practical safety, as noted in Table 2. In doing so, they elicit emotive responses such as "loneliness, anxiety, comfort."

Table 2: The mobile phone as source of reassuring (Survey: A Sociedade em Rede em Portugal[4] 2006, CIES-ISCTE)

While alone and waiting for someone, did you ever use your mobile phone		Male	Female
So no one would bother you	Yes	46%	54%
	No	51%	49%
To feel you are not alone	Yes	46%	54%
	No	51%	49%

There is a higher percentage of girls who use the mobile phone to feel they are not alone: 45.6% of girls compared with 39.3% of boys (Cardoso, Espanha et al., 2007). Women also feel calmer when they have their mobile phone with them: 49% of men against 51% of women (Cardoso, Gomes et al., 2007). Similarly, women tend to feel more anxious when they cannot have the mobile phone with them: 55% of women compared with 45% of men totally agree with this statement. Moreover 36.9% of girls use the mobile phone to avoid being bothered while that percentage drops to 21.8% for boys (Cardoso, Espanha et al., 2007, pp. 202–203). The theory of *Digital Agency* recognizes such emotions as fostering an individual's agency.

The mobile phone case study highlights the fragility of traditional stereotypes regarding women and technology but it also shows resistance points, the persistence of patriarchal views and the obstacles women face in technology appropriation. The

Five Acts of Agency provided a systematic framework of analysis to articulate the contradictions and tensions of the gendering process.

Concluding Remarks

The *Five Acts of Agency* has proven a useful framework with which to explore the everyday interactions women experience and the influences on their agency and ultimately participation. The application of the framework to two different empirical studies, their contexts and methods, highlight its strength as a unifying framework. The theory of *Digital Agency* is mainly about contextualized and situated acts and thus the choice to explore different contexts adds scope and breadth. This suggests that the *Five Acts of Agency* provides conceptual categories, which have the potential for application to a diverse range of situated or contextualized settings when exploring dynamic, emergent interaction. This being an important point when we recognize we live in an ever-changing society.

In the previous decade, concerns of a "digital divide" (Castells, 1996) led to a focus on an individual's equitable access to technology and information. Strategies to provide access in school settings proliferated, extending to flagship programs, such as "one laptop per child"[5]. Our current decade is characterized by ongoing rapid change. If one were to enter any suburban mall and observe people accessing their email on low-cost kiosks or their own laptops or "tweeting"[6] away on mobile phones, there is a sense that affordable access to technology is perhaps less of an influence on participation. However, as the case study in Portugal demonstrates, access does not translate into degendered practices or the end of women's alienation from technologies.

Although it was evident in the DCI case study that access to technology had enabled women's participation, particularly in their youth, several other influences were apparent. In drawing on the *Five Acts of Agency* as a framework, the case study illustrated that it is a diverse range of acts, which accumulate over a lifetime, that enable or constrain participation. For example, beyond access to technology, these women connected to technology through mastery experiences, developing a positive association with it through collaborating with others, and valuing the creative aspects of the industry. This led them to developing an emotional connection, a sense of passion, to their careers. The emerging theory of *Digital Agency* helps us to understand that women's participation, for example in using technology, is a sum of acts. Women can be enabled by some acts (as in parents actively providing young girls with mobile phones), or constrained by others (as in the stereotypes surrounding women's capability with technologies).

The theory of *Digital Agency* also brings a fresh perspective to the gender and technology debate, as it does not frame women as helpless victims. On the contrary, it sees them as aware agents whose acts can transform a circumstance. This contrasts with much of the previous research that simply adds to the long list of negative influences associated with women and their experiences with technology. Although, the emphasis in the reported studies has been on exploring the experiences of a gendered identity, the *Five Acts of Agency* framework may be useful in considering the influences on participation of other "minority" identities.

In conclusion, this paper has highlighted that the inequitable participation of women, particularly in areas involved with technology, can be explored using a human agency perspective. The exposition of the theoretical framework of *Digital Agency* and the insights from the empirical data, to some extent, helps answer the vexing question posed at the beginning of the chapter of how, and in what ways, the action of individual agents is related to the structural features of the society of which they are a part.

NOTES

1. This case study stems from a PhD funded by the Faculty of Science and Technology within Queensland University of Technology.
2. "m" refers to a participant from a multimedia organization and "g" refers to a participant from a games organization.
3. The respondents were between 9 and 18 years old. The data was gathered through an online questionnaire that aimed at studying the media consumption of young people in Portugal.
4. A national survey on the network society in Portugal based on a sample of the Continental Portuguese population, aged 15 or more.
5. The One Laptop Per Child Association, Inc. (OLPC) is a U.S. non-profit organization set up to oversee the creation of an affordable educational device for use in the developing world. See http://laptop.org/en/ (last accessed January 10, 2010)
6. Twitter (http://www.twitter.com) is a free social networking and micro blogging service that enables its users to send and read short messages known as tweets.

REFERENCES

Australian Interactive Media Industry Association (AIMIA) (2005) Digital Content Industry Roadmapping Study. Australia. Retrieved September 19, 2006 from http://www.aimia.com.au/i-cms_file?page=1455/Digital_Content_Roadmapping_Study_FINAL_AIMIA_Version.pdf
Archer, M. (1995). *Realist social theory: The morphogenetic approach.* Cambridge, UK: Cambridge University Press.
Archer, M. (2000). *Being human: The problem of Agency.* Cambridge, UK: Cambridge University Press.
Australian Bureau of Statistics (ABS) 2008. No. 4902.0. *Australian Culture and Leisure Classifications.* Canberra. Australia, Retrieved October 10, 2008 from http://www.abs.gov.au/AUSSTATS/abs@.nsf/DetailsPage/4902.02008%20(Second%20Edition)?OpenDocument.
Bandura, A. (1997). *Self-efficacy: The exercise of control.* New York: W. H. Freeman.

Bandura, A. (1999). Social Cognitive Theory: An agentic perspective. *Asian Journal of Social Psychology*, 2, 21–41.
Bandura, A. (2001). Social cognitive theory: An agentic perspective. *Annual Review of Psychology*, 52, 1–26.
Beekhuyzen, J.C., Clayton, K, (2004). Changing ICT career perceptions: Not so geeky? *1st International Conference on Research on Women in Information and Communication Technology*. 28–30 July, Kuala Lumpur, Malaysia.
Cardoso, G., Espanha, R., & Gomes, M. d. C. (2007). *E-Generation: Os usos de media pelas crianças e jovens em Portugal*. Lisbon: CIES ISCTE.
Cardoso, G., Gomes, M. d. C., Espanha, R., & Araújo, V. (2007). *Mobile Portugal*. Lisbon: Obercom.
Castells, M. (1996). *The rise of the network society*. Cambridge, UK: MA Blackwell.
Castells, M., Fernandez-Ardevol, M., Qiu, J., & Sey, A. (2007). *Mobile communication and society: A global perspective*. Cambridge,MA: The MIT Press.
Connell, R.W. (2002). *Gender*. Cambridge, UK: Polity.
Cunningham, S., Cutler, T., Ryan, M., Hearn, G., Keane, M. (2003) *Research and Innovation Systems in the Production of Digital Content and Applications. Content and Applications, Creative Industries Cluster Study Volume III*. Canberra: Commonwealth of Australia (DCITA) Retrieved November 18, 2009 from http://www.chass.org.au/papers/PAP20030901CG.php
DCITA (2005). *Australian digital content industry futures plan*. The Department of Communications, Information Technology and the Arts (DCITA), March 2005. Canberra: Commonwealth of Australia. Retrieved September 19, 2006 from http://www.dcita.gov.au/__data/assets/pdf_file/37474/Appendix_C.2_Australian_digital_content_futures.pdf
Deuze, M., Martin, C., & Allen, C. (2007). The professional identity of gameworkers. *Convergence*, 13(4), 335–353.
Eisenhardt, K. (1989). Building theories from case study research. *Academy of Management Review*. 14(4): 532–550.
Geneve, A., Nelson, K., & Christie, R. (2009). Women's participation in the Australian Digital Content Industry: initial case study findings. In Prpić, K ., Oliveira, L., Hemlin, S. (eds.) *Women in Science & Technology*. Zagreb, Croatia: Institute of Social Research-Zagreb Sociology of Science and Technology Network of ESA (European Sociological Association), 139–161.
Geser, H. (2004). *Towards a Sociological Theory of the Mobile Phone* [Electronic Version]. Retrieved February 22, 2007 from http://socio.ch/mobile/t_geser1.pdf.
Giddens, A. (1984). *The constitution of society*. Cambridge, UK: Polity.
Giddens, A (1993). *Sociology*. Cambridge, UK: Polity.
Giddens, A., & Turner, J. (1988). Analytical theorizing. In Giddens, A., Turner, J. (Eds.), *Social theory today*. Stanford, CA: Stanford University Press.
Gill, R. (2002). Cool, Creative and Egalitarian? Exploring Gender in Project-Based New Media Work. In *European Information, Communication and Society*, 5(1), 70–89.
Gill, R. (2007). *Technobohemians or the new Cybertariat? Understanding contemporary new media work*. Amsterdam: Institute of Network Cultures.
Griffiths, M., Keogh, C., Moore, K., Tattersall, A., & Richardson, H. (2005). Managing diversity or valuing diversity? Gender and the IT labour market. In Neiderman, F., Ferratt, T. (Eds.), *IT workers: Human capital issues in a knowledge-based environment*. Hershey, PA, USA: Information Science Publishing, 303–330.
Gürer, D., & Camp, T. (2002). *Investigating the incredible shrinking pipeline for women in computer science* (Final report). National Science Foundation Project 9812016. Retrieved May 6, 2006 from:

http://www.acm.org/women/ pipeline-finalreport_ver_2.doc

Haraway, D. (1991). *Simians, cyborgs and women: The reinvention of nature.* New York: Routledge.

Kvasny, L., & Trauth, E.M. (2002). "The 'Digital Divide' at Work and Home: Discourses about Power and Underrepresented Groups in the Information Society." In Wynn, E., Myers, M.D. and Whitley, E.A. (Eds.), *Global and Organizational Discourse about Information Technology.* Boston: Kluwer Academic Publishers: 273–291.

Lasen, A. (2004). Affective Technologies. Emotions and Mobile Phone.[Electronic Version]. *Receiver #11 Exchange.* Retrieved January 12, 2005 from http://www.receiver.vodafone.com/archive/index.html.

Lauretis, T. d. (2004). The technology of gender. In Rakow, L.F., Wackwitz, L. A. (Eds.), *Feminist communication theory: Selections in context.* (pp. 214–236). London: Sage.

Layder, D. (1998). *Sociological practice: Linking theory and social research.* London: Sage.

Lee, D. (2005). Women's creation of camera phone culture [Electronic Version]. *Fibreculture.* Retrieved February 9, 2008 from: http://journal.fibreculture.org/issue6/issue6_donghoo.html

Lemish, D., & Cohen, A. (2005). On the gendered nature of mobile phone culture in Israel. *Sex Roles: A Journal of Research, 52*(7–8), 511–521.

Ling, R. (2004). *The mobile connection: The Cell Phone's Impact on Society.* San Francisco, CA: Morgan Kaufmann.

Peril, L. (2002). *Think pink: Becoming a woman in many uneasy lessons.* New York and London: W. W. Norton & Company.

Plant, S. (2001). *On the mobile: The effects of mobile telephones on social and individual life*: Motorola. Retrieved December, 28, 2008 from http://www.motorola.com/mot/doc/0/234_MotDoc.pdf

Puwar, N. (2004). *Space invaders: Race, gender and bodies out of place.* Oxford and New York: Berg.

Quesenberry, J. (2006). Career anchors and organizational culture: A study of women in the IT workforce. In *Proceedings of the 2006 ACM SIGMIS CPR Conference on Computer Personnel Research: Forty Four Years of Computer Personnel Research: Achievements, Challenges & the Future,* April 13–15, 2006. Claremont, California, USA. New York: SIGMIS CPR '06. ACM, 342–344. Retrieved March 2, 2008 from: http://doi.acm.org/10.1145/1125170.1125249

Rakow, L., & Navarro, V. (1993). Remote mothering and the parallel shift: Women meet the cellular telephone. *Critical studies in mass communication, 20*(3), 144–157.

Ramsey, N., & McCorduck, P. (2005) *Where are the women in information technology? Report of literature search and interviews.* Prepared by the Anita Borg Institute for Women and Technology for the National Center for Women & Information Technology. National Science Foundation under Grant No. 0413538, Retrieved August 2, 2006 from http://www.anitaborg.org/news/publications/wherearethe_women.pdf

Riviere, J. (1929). Womanliness as masquerade. *The International Journal of Psychoanalysis, 10,* 303–313.

Roan, A., & Whitehouse, G. (2007). Women, information technology and "waves of optimism": Australian evidence on "mixed-skill" jobs. *New Technology Work and Employment, 22*(1), 21–33.

Skog, B. (2002). Mobiles and the Norwegian teen: Identity, gender, and class. In Katz, J. (Ed.), *Perpetual contact: Mobile communications, private talk, public performance.* Cambridge, UK: Cambridge University Press.

Stake, R. E. (1995). *The art of case study research.* Thousand Oaks, CA: Sage.

Steenson, M. (2006). *Mobile space is women's space: Reframing mobile phones and gender in an urban context.* Yale, CT: Yale School of Architecture.

Tashakkori, A., & Teddlie, C. (1998). *Mixed methodology: Combining qualitative and quantitative approaches.* Thousand Oaks, London & New Delhi: Sage.

Thompson, J. (1984). *Studies in the theory of ideology.* Cambridge, UK: Polity.

Trauth, E., Quesenberry, J., & Morgan, A. (2004). Understanding the under representation of women in IT: Toward a theory of individual differences. In *Proceedings of the ACM SIGMIS Computer Personnel Research Conference: Careers, culture, and ethics in a networked environment, SIGMIS'04,* April 22–24, 2004. Tucson. AZ, USA: ACM Press. Retrieved August 2, 2005 from http://portal.acm.org/citation.cfm?id=982372.982400&coll=GUIDE&dl=GUIDE&type=series&idx=SERIES303&part=series&WantType=Proceedings&title=CPR&CFID=://www.google.com/search?source=ig&CFTOKEN=www.google.com/search?source=ig

Trauth, E., Quesenberry, J., & Yeo, B. (2005). The influence of environmental context on women in the IT workforce. In *Proceedings of the 2005 ACM SIGMIS CPR Conference on Computer Personnel Research,* April 14—16, 2005. Atlanta, Georgia, USA. SIGMIS CPR '05. ACM, New York, 24–31. Retrieved February 8, 2007 from http://doi.acm.org/10.1145/1055973.1055979

Trauth, E.M., Howcroft, D. (2006). Critical research on gender and information systems. In Trauth, E.M (Ed.), *Encyclopedia of Gender and Information Technology.* 141–146.

Walsham, G. (1995). Interpretive case studies in information systems research: Nature and method. *European Journal of Information Systems, 4,* 74–81.

Yin, R. (1994). *Case study research: Design and methods* (2nd ed.). Beverly Hills, CA: Sage.

CHAPTER TWO

Snap, Post, Share

Understanding the Online Social Life of Personal Photography

ERIC COOK & CRISTINA GARDUÑO FREEMAN

In recent years there has been much discussion about the value and effect of amateur and individual media contribution online. But as Cheryl A. Casey describes, in her review (2007) of Andrew Keen's polemic *The Cult of the Amateur* (2007), this discussion is a case of "infinite monkeys versus the long tail, the professional versus the noble amateur." Popular books like Dan Tapscott's *Wikinomics* (2006) and Clay Shirky's *Here Comes Everybody* (2008) herald this wave as a normative good, one which is expanding civic participation, challenging the power of institutions, and changing accepted ways of working. Others, like Keen, argue that amateur content is a destructive cultural force. Using T.H. Huxley's evolutionary theorem to illustrate (where masterpieces are the result of infinite monkeys with infinite typewriters) he argues that "instead of creating masterpieces, these millions and millions of exuberant monkeys [networked individuals]—many with no more talent than our primate cousins—are creating an endless digital forest of mediocrity" (pp. 2–3). Keen's book is a response to Tapscott and Shirky's proposition that amateur content production will change the way we live. But all these propositions are trapped within the same paradigm: they judge amateur contributions against professional ones, rather than trying to understand these social activities on their own terms.

In this chapter, we consider everyday and personal media production, particularly those activities that centre on photography and photosharing. As cameras are integrated into digital devices, our rates of consumption and production of images is ever increasing, and photography is arguably becoming a ubiquitous practice of modern everyday life. More broadly, however, we are arguing against the simple dichotomous framing of such activities as amateur or professional, private or public, individual or collective. In view of this we adopt the term "personal photography" as used by media scholar Jose van Dijck (2008) to avoid the connotations usually associated with words like amateur, family, or tourist photography. New tools inevitably engender changes in the ways we interact, communicate, and record our everyday lives, and research is often quick to privilege the way technology *transforms* existing cultural practices. We take the position that often underlying and entwined with these new ways of doing things, are persistent conventions and habits supporting our ongoing social needs, *continuities* that reveal the complexities of social and visual interactions. In this chapter we address both narratives by examining how the practices of personal photography are maintained and transformed by the advent of websites like the online photo hosting sites Flickr and Picasa. In particular we understand that a photograph presents a different set of interpretive and evidentiary possibilities and characteristics than text. Photos are curiously polysemic: their meaning is ambiguous and specific at the same time (Mitchell, 2005; Chaplin, 2006), they are a medium of communication, objects of affect as well as locations, which support social interactions and moments of personal reflection. These uses of photography are not new, but their presence in the online domain is. Modern information and communication technologies (ICT) and increased access to digital media production tools have the effect of increasing the flexibility of these practices and extending these into new territories.

After locating our discussion within the broader literature on photography, we discuss the *use* and *extension* of two analytical perspectives used in the authors' own research: that of Chalfen's concept of the "home mode" of pictorial communication, and van Dijck's theory of "mediated memory." For each one, we *summarize* the core points of the authors' original argument and *illustrate* their application through examples drawn from our prior research. Each example considers personal photographs that are often described as mundane; a snapshot of a new home office, photos of food, a collection of almost identical views of a tourist landmark. We ask what compels individuals to take these photographs, how they are used, and what they might mean. We find that such photos, when viewed within their creator's social networks, have finely nuanced meanings contextualized by the relationship between the photographer and the viewer; posting an image of the home office can be a message to one's spouse that work is indeed being accomplished, whilst to a colleague an affirmation that the consulting venture is making progress. Similarly the photo of a landmark is a process of personal memory making as well as an artifact of social currency evidencing participation in larger social customs like tourism.

Through these examples, we highlight the strengths and weaknesses of each analytical perspective, demonstrate their applicability, and discuss how they supplement each other. Despite their differences, both perspectives support a key aspect of our argument that personal and everyday media must be analyzed in reference to their participants' goals, contexts, and behaviors, rather than treating the everyday media domain as subsidiary to dominant or professional cultural narratives. We suggest this not simply to replace over-privileging professional production with over-privileging amateur production. Instead, we seek to challenge researchers of technology to question their assumptions about the multiple functions of user-generated media—not to assume by default that all media production online inevitably targets markets (whether financial or reputational), and not to assume that individual contributions can be neatly assigned into boxes of meaning, interpretation, and valuation.

Personal Photography

Mobile phones, digital cameras, web cams; photography and the means for generating photographic images have proliferated in the last decade. Most people are now likely to own several image-making devices, such as a mobile phone with a camera, an integrated web camera on their computer, as well as a variety of digital cameras ranging from small compacts to semiprofessional models. But personal photography is not new. By the mid 1980s, 93% of US households owned a camera; each household produced an average of 126 photos each year (Chalfen, 1987, pp. 13–14.). The recent advent of more economical and available means of producing photographs has increased the pervasiveness and frequency of images as a mode of communication and a site of social interaction (Chalfen, 2006). This "pictorial turn" (Mitchell, 2005) has spurred academic interest in understanding how images communicate—within their own right—rather than with reference to text. Images, pictures, and photographs have specific qualities; they are affective objects, they require interpretation, and yet, they provide a sense of specificity.

For many years, much of the academic literature treated the photograph as either a fine art object (art history and criticism) or as a document in service of an event (photojournalism). Social scientists are also party to this limitation, focusing primarily on the object itself, rather than the actions and social interactions related to photography—producing them, displaying them, using them. Key exceptions to this generalization include Bourdieu's (1965/1990), Barthes' (1981), and Sontag's (1977) books on photography. Their canonical works provide important insights into some of the deeper implications of photography both as an activity and a medium. Scholars who have examined nonprofessional photography outside of online settings, such as cultural geographer John Urry and cultural studies academic Stuart Hall, evidence this approach.

Chalfen and the "Home Mode"

For our purposes we focus first on the work of anthropologist Richard Chalfen (1987). In the 1980s, Chalfen investigated the way home photography entailed more than the automated making of images, describing instead how these personal photographs serve to reinforce social relations. He termed this as the "home mode," a form of pictorial communication that supports "a pattern of interpersonal and small group communication centered around the home [...]. This concept of mode allows us to place pictures, as symbolic forms, into a process of social communication" (p. 8). Chalfen's goal was "to learn how people have organized themselves socially to produce personalized versions of their own life experiences [...] examining how a 'real world' gets transformed into a symbolic world" (p. 10). For Chalfen, "home" was intended to describe a social context, not just a geographical one. Home denotes the symbolic audiences of intimates, specifically addressing familial functions, in its production, usage, content, form, and functionality. Although there are many continuities in the way individuals use photographs in the "home mode," the advent of ICTs has greatly extended and developed their use; hence, we suggest that Chalfen's concept of photographic communication can be extended into the "virtual home mode," a concept we explore further in the section "Applying and Extending Chalfen"

Several findings of Chalfen's concept of the "home mode" are developed in our discussion. Chalfen found that these snapshots did not serve as stand-alone information objects; that is, as carriers of content or communication messages. The photos themselves are not creative visual stories or visual narratives. Rather, Chalfen showed that participants reconstructed narratives and interpretations, drawing on evidence and triggers found in the images. The photos serve as a *location* (literally and symbolically) for storytelling and memory construction. Seabrook (1991) reinforced this point in his examination of photo albums of working class UK families. As he notes, these albums do not tell a story directly, but rather they "illustrate a story" or "amplify biographies" (p. 172).

Next, Chalfen asserted that home mode media serves four "functional categories" in peoples' lives: documentary/evidentiary, preservation, memory, and cultural membership.

Home mode media provides data and *evidence* to construct and support familial stories. But this occurs selectively, via regularized patterns of inclusion and exclusion. Essentially, we document and create photographic evidence for the memories that serve and facilitate the retelling of those stories which serve our senses of self and of family; we tend to exclude representations that are locally irrelevant, unpleasant, or socially inappropriate. *Preservation* functions include the "capturing" and "encapsulation" of events and individuals, similar to Barthes' assertion (1981) that photography is a form of symbolic acquisition. *Memory functions* emphasize the photo as a location and locus for the "telling" and enactment of memories, rather than as a "container" for memories. *Cultural membership* functions operate at multiple levels of culture, from the very large to the local culture of the family and/or peer group. We both signify our cultural membership via our photographic practices and social uses, as well as have those practices and uses shaped in turn by our cultural membership, in what Chalfen labelled at the time as "Kodak Culture." These functional categories are not mutually exclusive—photos are polysemic both in their message and in their use.

Finally, Chalfen's work demonstrated how the home mode of symbolic production is distinct from commercial and fine art modes of photography. This difference is not just a question of skill or content, but types of audiences and social actions that are being addressed as well. To say that home mode producers are "doing it wrong" is a conceptual error on the part of the observer—it mistakes professional aesthetic standards as being the appropriate yardstick with the actual localized goals, meanings, and relevance of home mode activities.

van Dijck and "mediated memories"

Secondly, we ground our discussion through the more recent work of Jose van Dijck (2005; 2007; 2008). In her recent book *Mediated Memories* (2007) van Dijck turns her attention to understanding the continuities and transformations that ICTs afford to the practice of personal photography. She concurs with aspects of Chalfen's original theory of "amateur" or "personal" photography as a mode of interpersonal interaction. Her argument is that this is not new—photography has always "served as an instrument of communication and as a means of sharing experience" (2008, p. 59). But she further expands the boundaries of these interactions and asks; "what is *personal* cultural memory and how does it relate to collective identity and memory?" (2007, p. 1, italics in the text). Whilst Chalfen's theory is bounded by the intimate and familial, van Dijck moves beyond this to consider the broader implications and relationships of these cultural practices both on an individual and collective level, and from a technological and cultural perspective. Thus, she proposes the concept of "mediated memories" as a way of accounting for the interdependent connection between "personal collections and collectivity but also to help theorize the *mutual shaping* of memory and media" (2007, p. 2, italics in text). The concept of mediated memories provides a way to understand the practices outlined by Chalfen in a wider social context—one that is less constrained by disciplinary boundaries. Chalfen's theory helps us to understand the localized implications of technology to personal photography, whilst van Djick provides a framework within which to analyze the way the personal and the collective are co-constituted.

Van Dijck begins by exploring *personal cultural memory* and the way it has been the domain of specific disciplines; namely psychology and neuroscience. She asserts that autobiographical memory is crucial to an individual's sense of self. Without memories we would have no sense of the past or the future—no sense of continuity. But she emphasizes the way certain rituals—like taking a photo of a landmark or making a record of a holiday meal—are the result of 'culturally agreed' ways of constructing memory, and acknowledges the immense pleasure many derive from creating these personal histories.

She then addresses the dominance of sociologists, historians, and cultural theorists in discussing *collective cultural memory*. Her critique of these disciplines' existing conceptions is that they do not sufficiently acknowledge the way collective notions are structured by individual experiences. Further, they also dismiss the way technologies are an integral part of the "making" of memory. Diaries, photos, videos—all enable memories to be created in specific ways. But we also imagine our lives through these technologies; we describe it as like being in a film, or a "hallmark" moment. But these artifacts are often conflated with the experiences they seek to preserve. Photos and similar memory media are seen as "containers," yet they simultaneously threaten the purity of remembrance. To illustrate, consider the disdain held for tourists who spend their holidays, camera pressed to the eye, versus the "nobler" traveler who consciously experiences without a camera "to remember." For van Dijck "memory is as much about the privacy to inscribe memories for oneself and the desire to share them only with designated recipients as it is about publicness, or the inclination to share experiences with a number of unknown viewers or readers." Van Dijck's perspective on these socio-cultural processes opens up discussion that bridges these realms.

In the two sections that follow, we make use of Chalfen's and van Dijck's analytic perspectives to shed light on online photosharing practices. What happens when photography enters the online realm? ICTs and digitization enhance some factors of a photograph: the opportunities for editing and distributing. "Digitization is often considered the culprit of photography's growing unreliability as a tool for remembrance; but in fact, history shows the camera has never been a dependable aid for storing memories" (van Dijck, 2007, p. 99). Photographs have always been subject to visual retouching and having the memories and experiences they communicate edited in recollection. This supports the need to consider personal photography, not simply as a concrete object, but rather as a practice or activity, where the artifact is understood in relation to its production and consumption.

On sites like Flickr and Picasa the personal and the collective meet: members have control over their images and how they are displayed to their familial networks, but at the same time they can make their images publicly accessible, or on Flickr, add them to collectives called "groups." In the section "Applying and Extending van

Dijck: Interpreting Collective Memory Through Flickr Groups," Garduño Freeman observes the online activites of one such group, formed to collect photographs of the Sydney Opera House, a well-known tourist landmark. The photographs and the online activities that surround them are opportunities to derive important cultural knowledge and to begin to understand how the personal and the collective are mutually shaping each other. In contrast, Cook (in the section "Applying and Extending Chalfen") explores through interviews the nuanced personal meanings and communications that photographs through online systems can provide. The study extends Chalfen's "home mode" and investigates the way these media technologies are shaping our intimate social interactions.

Applying and Extending Chalfen

The concept of the home mode has been criticized as "ahistorical" by those who favor more ideologically deterministic explanations of personal media (Zimmerman, 1995). However, a closer reading underscores the usefulness of the home mode concept, which can be shown to be flexible enough to accommodate changes in production and distribution technologies, as well as in family structure and ideologies (Moran, 2002). In the overlapping fields of Human-Computer Interaction, Computer-Supported Cooperative Work, Computer-Mediated Communication (CMC), and Social Computing, Chalfen's work has gained renewed attention amongst researchers interested in photographic practice and sociality mediated via photography, having been cited in studies such as Miller and Edwards (2007), Frohlich, Kuchinsky, Pering, Don, and Ariss (2002), Van House et al. (2004, 2005), and Van House (2007). Other related work does not directly reference the home mode, but clearly addresses a similar set of activities, participants and functions, such as research on the camera phone by Kindberg, Spasojevic, Fleck, and Sellen (2005), Ling (2008), Ito (2005), and Okabe and Ito (2006). This literature argues for the ongoing analytic value of the home mode concept, as well as underscoring the need to continue updating and revising it. We must consider the ways that modern ICT and CMC may be supporting and changing the home mode, as it moves increasingly from the living room to networked communications; we must develop a model of the *virtual home mode* (VHM).

Personal Photography Practices and the Virtual Home Mode

In 2008 and 2009, Cook examined the relationships between everyday digital photography practices and personal well-being. As part of this study, he interviewed 24 individuals in their homes or workplaces. The participants included 10 men and 14 women, recruited across five distinct life stages: single young adults, married without children, married with children, "empty nest" adults, and elders. Participants were selected via purposeful sampling, recruiting individuals who had engaged in virtual home mode activities (sharing photos online with family and/or friends) regularly for at least a year. This sampling approach was activity-focused, rather than system-specific. As a result, the participants reported using a wide variety of tools, services, and approaches for sharing their photos. These included, but were not limited to: Snapfish, Shutterfly, Blogger, Facebook, personal websites, Kodak gallery, Livejournal, Photo.net, Picasa, Apple MobileMe, Flickr, Yahoo groups, private email lists and, in two cases, self-coded photo management systems written in PHP, HTML, and Perl. The participants often reported the use of multiple systems, contingent on their target audience and the intended function of their VHM activities.

The primary data for this study consisted of semi-structured interviews and observation sessions. Conducting the sessions face-to-face in their home or work spaces allowed the interviews to be structured primarily through a series of photo-elicitation tasks, using the participant's own photographs to contextualize and focus their responses. In addition to revealing general photography practices, this photo-elicitation protocol highlighted decisions related to media production and sharing, as well as prompting reflection on recent life events, and the photographic representations of those life events. Centering the research in the home revealed local context and personal meanings that would not have been otherwise visible to a researcher. Similarly, this approach also provided access into the patterns of exclusion, revealing not only what images had been posted publicly, but also those that were private or never shared at all.

This study used key concepts from Chalfen's home mode model to direct both the semi-structured interview protocol and the iterative qualitative coding of transcripts. The application of the home mode lens proved invaluable in understanding individual photographer's accounts and actions, as well as emphasizing the need to extend it for current socio-technical settings and to develop the concept of the virtual home mode. The following examples illustrate two key points: the necessity of the photographer's local context for full interpretation and the way ICTs allow new audiences to connect with traditional home mode representation.

"The traditional foods"

One of the more commonly used examples of the banality of online amateur photography is the food photograph. "Who cares what you had for breakfast?" is the standard refrain. Yet viewed through the lens of the virtual home mode, a seemingly banal food photo can reveal itself to be detailed and nuanced.

Figure 1. Hanukkah table setting. *Image reproduced with permission from study participant.*

An example of this presented itself during a June 2009 interview with "Jody," a professional woman in her late twenties whose family lived several states away from her. During the course of the interview, Jody presented many photos of food, discussing their use in her related hobbies of cooking and baking. Particularly salient was a photo of a holiday dinner setting that she had posted via Picasa. Jody explained the story behind this image, stating "so I celebrate Hanukkah and I had a little party here and so I posted for my dad because he was sad that I wasn't home to celebrate with the family. [The picture of the food at dinner] said don't worry, I'm still celebrating. [...] I'm still making the traditional foods over here, don't worry."

A simple documentary photo of a holiday dinner takes on new significance when the context of the VHM participants is considered. In this quote, we see representational family communication combined with the personal symbolism that exists both in personal photos and home-cooked food. Maintenance of distant family ties are wrapped together with adherence to cultural and religious traditions, and these myriad activities are managed in a single photo posted online. This example also underscores the challenge of reading and interpreting the meaning and value of a VHM photo by an uninvolved outsider. Jody's food image is not a self-contained message, but in a fashion similar to Brown and Duguid's analysis (1996) of the "social life of documents," instead supports creating common ground and shared awareness of activity. In so doing, the photo may "underwrite social interactions; not simply to communicate, but also to coordinate social practice" (ibid., p. 3).

"The home office"

Participants in the home mode share a history, imbuing images with additional context and meaning beyond what is immediately visible. One new but key aspect of the VHM is that multiple audiences can be granted access to the same photo. In multiple interviews during this study, Cook saw examples of photographers intentionally addressing multiple audiences concurrently, through the careful awareness of the particular frames of interpretation that will be brought to bear on a given image by different viewers.

Consider the example of study participant "Bob," a 36-year-old man who posted a picture of his new home office on his Flickr account. This participant had recently lost his job, and had begun to do freelance consulting. In Bob's words, the office image was "doing multiple things at once... like a good book." The messages that he intended to convey depended on the audience viewing the image. For geographically distant family members, the image was a message that he was coping emotionally with the loss of his job, and moving forward. For local professional and casual friend contacts, the home office image was to be interpreted in conjunction with other images Bob posted around the same time, showing activities such as trade luncheons and industry workshops that he would not previously have had time to attend. In Bob's account, these images were public signals that he was available but also still professionally active, without having to explicitly state that he was unemployed.

Bob labeled this as "sideways" maintenance; he was able to send distinct but related signals to both of these audiences at the same time, but without the social embarrassment of having to address the topic head-on. He engaged in home mode style photo production, producing a seemingly mundane/photo with no intent of being professional and aesthetically "good," which served communicative purposes and leveraged previous shared histories between producer and audience for interpretation. Yet simultaneously, his photo was serving professional and communicative functions for non-home mode audiences. Bob's office photo was an intentionally constructed boundary object (Bowker and Star, 1999), crossing between multiple social spheres and coordinating with each in a meaningfully distinct fashion. As we will see in the following section, the boundary object aspect of personal photography takes on an additional perspective via van Dijck's work.

Applying and Extending van Dijck: Interpreting Collective Memory Through Flickr Groups

Flickr is often cited as an example of the Web 2.0 participatory turn, and has already generated a substantial body of research. Most of these studies are empirical and take advantage of Flickr's open source platform and metadata to analyze the site's social networks (Lerman & Jones, 2007; Lerman, 2007; van Zwol, 2007;

Sigurbjörnsson & van Zwol, 2008), classification systems—folksonomies (Kennedy et al., 2007; Davies, 2006; Rafferty & Hidderley, 2007; Yakel, 2006; Lerman et al., 2007), socio-locative practices (Ames & Naaman, 2007; Erickson, 2007), groups (Negoescu & Gatica-Perez, 2008; Pissard & Prieur, 2007) and as an archive of digital photographs (Van House, 2006; 2007; Van House & Ames, 2007; Van House & Churchill, 2008). Although these empirical studies provide information about usage patterns like the number of images contributed and social networks between members or participation in groups, they do not analyze the photographs themselves and the way group members negotiate their collective identities through visual and cultural means. Two exceptions are Jean Burgess' (2007) thesis, which explores how Flickr is a civic space for enactments of vernacular creativity and cultural citizenship, and Janice Affleck's (2007) investigation of the opportunities that spaces like Flickr provide for the discursive interpretation of heritage by communities. The example described here proposes that Flickr is a space where personal photographs are a medium through which public memory about iconic architecture is constructed. Arguably, this is a key transformation afforded by ICTs: not that these activities themselves are new, but that these discussions take place in public, that individuals negotiate via visual representations and that these collections of photographs give rise to new public formations.

Flickr Group: "Sydney Opera House"

During 2008 and 2009 Garduño Freeman observed the social interactions of one specific Flickr group called "Sydney Opera House." The group's focus is gathering photographs of Sydney's famed landmark, ones that position the building as the main subject and prominently within the frame. Using the Sydney Opera House group as the filter for the images reveals the way collective identity is negotiated through personal photography and photosharing on Flickr. These exchanges are "visual conversations" where members negotiate their personal identity; at the same time these technologies and their respective cultural practices simultaneously shape these conversations. Garduño Freeman's observations reveal the way the representation and our *collective cultural memory* of this building are intricately connected. Ultimately the study seeks to understand the role of the Sydney Opera House, as a World Heritage site, in the everyday life of its communities. The visual and textual discussions contributed by members of this group demonstrate the complex and layered sentiment held for this place.

The Sydney Opera House is widely accepted as a masterpiece of expressive modern architecture of the twentieth century. It is acclaimed for its form and structural innovation. In June 2007, the Sydney Opera House was finally inscribed onto the UNESCO World Heritage list, described as "a great urban sculpture set in a remarkable waterscape." The listing recognizes "the building has had an enduring influence on architecture." (UNESCO, 2007). Yet this building, less than half a century old, has been surrounded by controversy for much of its life. The design is the winning entry of an international competition held in 1957, by Danish architect Jørn Utzon. Utzon's nontraditional working methods and an unrealistic construction schedule coupled with Australian politics during that time led to an unworkable relationship (Murray, 2004; Drew, 2001). After Utzon's resignation in 1966, the building was completed by local architects Peter Hall, Lionel Todd, and David Littlemore, a turn of events that has divided the local Sydney community and architects abroad. The recent inscription as a World Heritage site, along with the award of the Pritzker Prize for Architecture in 2003 has been seen as the reconciliation of Australia with Utzon. The Sydney Opera House continues to be an object of much debate and affection, both as an object in itself and as a recognized symbol for Sydney.

Figure 2. Thumbnail mode showing group pool from "Sydney Opera House" http://www.flickr.com/groups/sydneyoperahouse/pool/

Photographs courtesy of "Sydney Opera House" group members (from top left to right) Marc Emond, Peter Lee, Lee Gilbert, Laurie Wilson, Ben Hockman, Alastair McAlpine, "L_Plater," Pascal Bovet, "scott_aus," Nick Barta, Carlos Lopez Molina, Laurie Wilson, "fotografX.org," "allrose," Ben Ward, Peter Lee, "flgirlinsydney" and Peter Lee. Image reproduced with permission of Yahoo! Inc. ©2009 Yahoo! Inc. FLICKR and the FLICKR logo are registered trademarks of Yahoo! Inc.

A search for Sydney Opera House on Flickr returns 123,000 images in January 2010. These are publically posted photographs on Flickr, many of which appear at first to be mundane tourist shots with little to contribute to the cultural knowledge of this place. However, tourist photographs are personal memories and expressions, and on Flickr are also contributed to groups. Groups are common spaces on Flickr where members can contribute photographs on subjects of common interest. Groups have specific curatorial guidelines; and observations show that photographs of the Sydney Opera House are contributed to various of these. Contribution to groups is one way that people make concrete their encounters with this building. The case study undertaken on the Sydney Opera House demonstrates the way members interact through and around these images via the possibilities afforded by websites like Flickr. Through "groups," Flickr members self-organize and negotiate a dynamic and fluid collective identity as well as furthering the formation of other groups, and defining the types of social patterns which occur on this site.

At the time of writing (January 2010), the Flickr group "Sydney Opera House" has over 840 members and some 2800 photographs in its "pool," varying from highly skilled photographs to low quality snapshots (see Figure 1). Flickr groups might easily be equated to an exhibition space or a photographic archive. But groups are more complex than simply a place to display or contribute photographs. Groups are social structures governed by administrators at the highest level and members at the lowest level. Like a club or any organization, groups are subject to peer pressure and social dynamics—some members are highly vocal in discussion threads, whilst others are avid contributors to the photographic pool. The socio-visual interactions in the Sydney Opera House group demonstrate the way its formation is tied to the high contribution of images of this place to more general Flickr groups like "Sydney, Australia." The dominance of photos of this building in "Sydney, Australia" is the trigger for the formation of both "Sydney Opera House" and its counterpart "Sydney-alt," a group defined by the exclusion of images of this building. This connection can only be discerned by visual analysis of the photographs contributed, in conjunction with the group members' discussion threads. Unlike much of the current research on Flickr, which provides data about the number of images tagged, or the number of members who participate in a group, close visual analysis can account for the way the visual representation of a landmark or tourist destination affects group member's sense of personal and collective identity. Here members contribute personal expressions, their photographs, and through them gain membership to a space in which cultural identity and memory can be actively negotiated.

Social media like Flickr are not just the territory of information science. Flickr and the photosharing practices it supports provide new opportunities for understanding both individual and collective sentiment about almost any subject. Taking photographs of the Sydney Opera House is not new, but using these images to participate in cultural negotiations about this place is a key transformation afforded by this technology. For the study of this building's social significance, within the frame of Heritage studies, Flickr groups like "Sydney Opera House" reveal new kinds of public engagements with this building. More broadly it demonstrates that photosharing on Flickr is a public visual discourse, a discursive practice involving interpretation and negotiation of personal and cultural identity. It demonstrates van Dijck's proposition; personal expression is tied to personal cultural identity; it is not merely a building block of the collective cultural memory, but rather shapes cultural memory and in turn is shaped by it. The photographs of the Sydney Opera House contributed to the group are inevitably influenced by professional representations of this place. But their sheer volume on sites like Flickr, regardless of their "comparable quality," serves to democratize the representation of this building. Further, the negotiations that occur in Flickr groups fracture the consensual notion that public sentiment towards the Sydney Opera House is straightforward; it is revealed as complex, multivalent, and generative. On Flickr the Sydney Opera House does not simply stand in symbolically for Sydney and Australia; but rather it is co-constituted as a locus, a meaningful place in the lives of its contemporary communities.

DISCUSSION AND REFLECTION

Perspectives such as those described above are intended to broaden the understanding of vernacular photographic practices. Cook's research focuses on developing the concept of the virtual home mode, whilst Garduño Freeman's case focuses on using photosharing online to understand collective memory. These agendas are not in conflict, but rather private and public perspectives of the same phenomenon. Cook's research has involved gaining access to private interviews with photographers to understand the way an image is intended for multiple audiences whilst Garduño Freeman has focused her observations on the public interpretations of images through groups. By considering these two approaches we can see that ordinary photographs are consequential. They need to be revealed as meaningful, both in individual and collective communication, and within existing and newly generated social networks.

Moreover, the approaches we have described above emphasize that personal media should be *approached on their own terms*, rather than viewed simply as unskilled or incompetent versions of professional production. Personal photographs may be visually pleasing or blurry and ill lit, poorly executed or masterful. Yet for the participants, they serve a purpose regardless. A point of connection may be drawn with Becker's descriptions of folk art activities in his broader analysis of "art worlds" (1982). Becker noted that for many instances of folk art, the social cohesion function of engaging in the act of artistic production is often more important than the quality of an art object being produced. One example he provides is the occasion of singing the traditional "happy birthday" song. While singing is a form of artistic expression that can involve years of training, precise technique and a detailed set of professional practices, these aspects are not important in the setting of a birthday celebration. Rather, what is important is that the social functions supported by singing are served, rather then the aesthetic qualities of the output. Put another way, what often matters most in this context is not that it gets done *well*, but that it gets done at all.

Ultimately, concerns of quality in personal photography are less relevant than the meanings and operations that such media serve for their constituents. In the examples provided from Cook's research, we saw the manner in which particular photos are positioned at the intersection between overlapping social spheres, and how they leverage shared context to build meaning and significance. In the example of Sydney Opera House research, Garduño Freeman notes that these Flickr images are uncommissioned, and that the discussions within the Flickr groups are self-organized. Instead of making broad generalizations about the way the public relates to this iconic location, we see evidence of public negotiation of meaning, the processes of mediation between individual and collective memory revealed. In each case, assessing personal photography and photographic practices on its own terms—that is, within the socially localized spheres of use and evaluation—was necessary to reveal hidden meaning and function.

How should those who wish to understand personal photography (online or off) on its own terms approach the subject? As researchers, we must first be cautious of assigning meaning to photographs, superimposing our assumptions or ideological stances, without first checking them against the local social reality of the photographic practitioners. We may find that our readings are valid, but they must be extracted from the participants' actions and accounts. We can seek out their intentions and interpretations in many ways. In Cook's work, this occurs by context unveiled through personal interviews and image elicitation methods; Garduño Freeman observes interactions revealed through behavioral traces left behind in the online setting of Flickr. Other approaches clearly exist; they will be connected by their desire to start by respecting the agency of the personal photographer.

Approaching personal media on its own terms also argues against deterministic or dominant use model conceptualizations of social media, which treat *system* and *community* as equivalent concepts. Our illustrations show that universal notions that make equivalences between systems and communities are often over-simplications. There is not one homogeneous "community of Flickr" or one "community of Youtube"; rather, there are multiple concurrent communities, co-habitating in *infrastructures* of production and dissemination. This concurs with Michael Warner's *Publics and Counterpublics* (2002) where he addresses the notion that publics are complex and multifarious entities. Warner disagrees with much of the literature in the social sciences, which frames publics as existing entities to be studied empirically. Warner proposes a more interpretive approach towards publics, one that embraces these social entities as animated, dynamic, and multileveled (Loizidou, 2003, p. 77). Further he argues that rather than producing texts, publics emerge in relation to texts; each "text" (or photo or Flickr group) co-constitutes an audience, and a public:

> "Each time we address a public (...), we draw on what seems like simple common sense. If we did not have a practical sense of what publics are, if we could not unself-consciously take them for granted as really existing and addressable social entities, we could not produce most of the books or films or broadcasts or journals that make up so much of our culture; we could not conduct elections or indeed imagine ourselves as members of nations or movements. Yet publics exist only by virtue of their imagining. They are a kind of fiction that has taken on life, and very potent life at that." (Warner 2002, p. 8)

These multiple and more specific conceptions of the value of personal photography have greater implications; as a research area it connects disparate academic disciplines and builds programs of research, rather than enforcing a specialist model. This offers the opportunity to connect research more effectively across different sociotechnical contexts. And indeed, this may be not only valuable but also required, as technological changes force different academic disciplines into new configurations, confronting new interdisciplinary problems. By crossing traditional boundaries, we can reconsider behaviors and relationships that are re-instantiated in each new generation of technology, and more readily highlight the permutations that new technology co-constructs. By treating each generation of system as new locations of inquiry into more persistent and underlying behaviors/phenomena, rather than as isolated and ahistorical cases, we may begin to paint a richer picture of technology and sociality, publics and memory. Our assumptions—as researchers and as photographic participants—may be challenged, and we may find that we have much to learn in the process.

References

Affleck, J. (2007). *Memory capsules: Discursive interpretation of cultural heritage through digital media.* Unpublished doctoral dissertation. The University of Hong Kong. Hong Kong, China.

Ames, M., & Naaman, M. (2007, April 28—May 3). Why we tag: Motivations for annotation in mobile and online media. *Proceedings of the SIGCHI conference on Human factors in computing systems*, San Jose, California, USA. Retrieved November 21, 2008, from http://portal.acm.org/citation.cfm?id=1240624.1240772

Barthes, R. (1981). *Camera lucida: Reflections on photography.* (trans. R. Howard). New York: Hill and

Wang.
Becker, H. S. (1982). *Art worlds*. Berkeley, CA: University of California Press.
Bourdieu, P., with Boltanski, L., et al. (1965/1990). *Photography: A middlebrow art.* (trans. Whiteside, S.). Stanford, CA: Stanford University Press.
Bowker, G., & Star, S. L. (1999). *Sorting things out: Classification and its consequences.* Cambridge, MA: MIT Press.
Brown, J. S., & Duguid, P. (1996, May 6). The social life of documents. *First Monday.* 1(1). Retrieved from http://www.firstmonday.dk/issues/issue1/documents/
Burgess, Jean, (2007). *Vernacular creativity and new media.* Unpublished doctoral dissertation. Creative Industries Faculty, Queensland University of Technology, Brisbane, Australia.
Casey, C. A. (2007, 22 September). The cult of the amateur. *Academic Commons.* Retrieved February 22, 2010, from http://www.academiccommons.org/commons/review/cult-of-the-amateur
Chalfen, R. (1987). *Snapshot versions of life.* Bowling Green, OH: Bowling Green State University Popular Press.
Chalfen, R. (2006). "Can you see me now?" Problems in the study of camera-phone use in the U.S. and Japan. *Proceedings of the International Visual Sociology Association Conference 2006. Urbino, Italy.* Retrieved from http://www.visualsociology.org/proceedings/proceedings_2006/Chalfen_see_me_now.doc
Chaplin, E. (2006).'The convention of captioning: W. G. Sebald and the release of the captive image.' *Visual Studies, 21*(1), 42–53.
Davies, J. (2006). Affinities and beyond! Developing ways of seeing in online spaces. *E–Learning, 3*(2). Retrieved November 21, 2008, from http://www.wwwords.co.uk/pdf/freetoview.asp?j=elea&vol=3&issue=2&year=2006&article=8_davies_elea_3_2_web/
Drew, P. (2001). *The masterpiece, Jørn Utzon: A secret life.* Victoria, Australia: Hardie Grant.
Erikson, I. (2007). Understanding socio-locative practices, In *Proceedings of Conference on Supporting Group Work (GROUP'07) Doctoral consortium papers* [online], Sanibel Island, Florida, USA, November 4–7. Retrieved November 21, 2009, from: http://portal.acm.org/citation.cfm?id=1329113
Frohlich, D., Kuchinsky, A., Pering, C., Don, A., & Ariss, S. (2002, November 16–20). Requirements for photoware. In *Proceedings of the 2002 ACM Conference on Computer Supported Cooperative Work* (New Orleans, Louisiana, USA, November 16—20, 2002). *CSCW '02.* ACM, New York, 166–175.
Ito, M. (2005, September 11–14). Intimate visual co-presence. *UbiComp 2005*, Takanawa Prince Hotel, Tokyo, Japan.
Keen, A. (2007). *The cult of the amateur: How today's internet is killing our culture.* New York: Doubleday/Currency.
Kennedy, L., Naaman, M., Ahern, S., Nair, R., & Rattenbury, T. (2007, Sep 23-28). How Flickr helps us make sense of the world: Context and content in community-contributed media collection. In *Proceedings of the 15th international conference on Multimedia (MM'07)*, Augsburg, Bavaria, Germany. Retrieved 21 November, 2008 from http://portal.acm.org/citation.cfm?id=1291384
Kindberg, T., Spasojevic, M., Fleck, R., & Sellen, A. (2005). "I saw this and thought of you: Some social uses of camera phones." *CHI '05* Extended Abstracts on Human Factors in Computing Systems, Portland, OR.
Kindberg, T., Spasojevic, M., Fleck, R., & Sellen, A. (2005). "The ubiquitous camera: An in-depth study of camera phone use." *Pervasive Computing, IEEE , 4*(2), 42–50.
Lerman, K. (2007, October 30). Social browsing and information filtering in social media, *Cornell*

University Library arXiv.org, arXiv:0710.5697v1 [cs.CY]. Retrieved November 21, 2008, from http://arxiv.org/abs/0710.5697v1

Lerman, K., & Jones, L.A. (2007). Social browsing on Flickr. *Proceedings of International Conference on Weblogs and Social Media (ICWSM'07)*. Boulder, Colorado, USA. Retrieved November 21, 2008, from http://www.icwsm.org/papers/3—Lerman-Jones.pdf

Lerman, K., Plangprasopchok, A., & Wong, C. (2007, April 12). Personalizing Image Search Results on Flickr, *Cornell University Library arXiv.org*, arXiv:0704.1676v1 [cs.IR]. Retrieved November 21, 2008, from http://arxiv.org/abs/0704.1676v1

Ling, R. (2008). *New tech, new ties*. Cambridge, MA: The MIT Press.

Loizidou, E. (2003). Book Review: Publics and Counterpublics, by M. Warner (2002). *Space and Culture*, 6(1), 77-78.

Miller, A. D., & Edwards, W. K. (2007, April 28–May 3). Give and take: A study of consumer photo-sharing culture and practice. *CHI 2007*, San Jose, California, USA. Retrieved July 29, 2009, from http://www.cc.gatech.edu/~keith/pubs/chi2007-photosharing.pdf

Mitchell, W. J. T. (2005). *What do pictures want?* Chicago and London: The University of Chicago Press.

Moran, J. M. (2002). *There's no place like home video*. Minneapolis, MN: University of Minnesota Press.

Murray, P. (2004). *The saga of the Sydney Opera House*. London and New York: Spon.

Negoescu, R., & Gatica-Perez, D. (2008, July 7–9). Analyzing Flickr groups. *Proceedings of the 2008 international conference on Content-based image and video retrieval (CIVR'08)*, Niagara Falls, Ontario, Canada. Retrieved July 29, 2009, from http://portal.acm.org/citation.cfm?id=1386406

Okabe, D., & Ito, M. (2006). Everyday contexts of camera phone use: Steps toward technosocial ethnographic frameworks. In J.R. Hoflich, & H. Maren (Eds.), *Mobile communication in everday life: Ethnographic views, observations, and reflections*. Berlin, Germany: Frank & Timme GmBh.

Pissard, N., & Prieur, C. (2007). Thematic vs. social networks in web 2.0 communities: A case study on Flickr groups. *Institut National de Recherche en Informatique et en Automatique (INRIA) HAL-CCSD*. Retrieved November 21, 2008, from http://hal.inria.fr/docs/00/17/69/54/PDF/42-algo-tel-flickr.pdf

Rafferty, P., & Hidderley, R. (2007). Flickr and democratic indexing: dialogic approaches to indexing, *Aslib Proceedings: New Information Perspectives*, 59(4/5), pp. 397–410. Retrieved November 21, 2008, from http://www.emeraldinsight.com/Insight/viewPDF.jsp?contentType=Article&Filename=html/Output/Published/EmeraldFullTextArticle/Pdf/2760590407.pdf

Seabrook, J. (1991). My life is in that box. In J. Spense & P. Holland (Eds.), *Family snaps: The meanings of domestic photography*. London: Virago.

Shirky, C. (2008). *Here comes everybody: The power of organizing without organizations*. New York: Penguin.

Sigurbjörnsson, B., & van Zwol, R. (2008, April 21–25). Flickr tag recommendation based on collective knowledge. *Proceedings of the 17th international conference on World Wide Web (WWW)*, Beijing, China. Retrieved November 21, 2008, from http://portal.acm.org/citation.cfm?id=1367542

Sontag, S. (1977). *On photography*. New York: Farrar, Straus and Giroux.

Sydney, Australia, (2005–2009). *Flickr*. Retrieved 2007–2009 from http://www.flickr.com/groups/sydneyaustralia/

Sydney-alt, (2006–2009). *Flickr*. Retrieved 2007–2009 from http://www.flickr.com/groups/33606562@N00/

Sydney Opera House, (2006–2009). *Flickr*. Retrieved 2007–2009 from http://www.flickr.com/groups/sydneyoperahouse/

Tapscott, D. & Williams, A. (2006). *Wikinomics: How mass collaboration changes everything*. New York, NY: Portfolio.
UNESCO (2007). Sydney Opera House, *UNESCO World Heritage Website*. Retrieved February 27, 2010 from http://whc.unesco.org/en/list/166
van Dijck, J. (2005). From shoebox to perfromative agent: the computer as personal memory machine. *New Media and Society*. 7(3), 311–331. Retrieved July 29, 2009, from http://nms.sagepub.com/cgi/content/abstract/7/3/311
van Dijck, J. (2007). *Mediated memories in the digital age*. Stanford, CA: Stanford University Press.
van Dijck, J. (2008). Digital photography: Communication, identity, memory. *Visual communication*, 7(1), 57–76. Retrieved July 29, 2009, from http://vcj.sagepub.com/cgi/content/abstract/7/1/57
Van House, N., Davis, M., Takhteyev, Y., Good, N., Wilhelm, N., & Finn, M. (2004, Nov 6–10). From "What?" to "Why?": The social uses of personal photos. In *CSCW'04*, Chicago, Illinois, USA. Retrieved July 29, 2009, from http://people.ischool.berkeley.edu/~vanhouse/van%20house_et_al_2004a.pdf
Van House, N., Davis, M., Ames, M., & Finn, M. (2005). The uses of personal networked digital imaging: an empirical study of cameraphone photos and sharing. *Proceedings of ACM CHI'05, Conference on Human Factors in Computing Systems, 2005*.
Van House, N. (2006). Interview viz: Visualization-assisted photo elicitation, in *Conference on Human Factors in Computing Systems CHI*, Montreal, Quebec, Canada. Retrieved November 21, 2008, from http://portal.acm.org/citation.cfm?id=1125451.1125720
Van House, N. (2007, April 28–May 3). Flickr and public images-sharing: Distant closeness and photo exhibition' *Conference on Human Factors in Computing Systems CHI '07*, San Jose, California, USA. Retrieved November 21, 2008, from http://people.ischool.berkeley.edu/~vanhouse/VanHouseFlickrDistantCHI07.pdf
Van House, N., & Ames, M. (2007, April 28–May 3). The Social Life of Cameraphone Images, paper under review for *Conference on Human Factors in Computing Systems CHI '07*, San Jose, California, USA. Retrieved November 21, 2008, from http://people.ischool.berkeley.edu/~vanhouse/photo_project/pubs/Van%20House%20and%20Ames.pdf
Van House, N., & Churchill, E. (2008). Technologies of memory: Key issues and critical perspectives, *Memory Studies*, 1(295). Retrieved November 21, 2008, from http://mss.sagepub.com/cgi/content/abstract/1/3/295
van Zwol, R. (2007). Flickr: Who is looking? In *2007 IEEE/WIC/ACM International Conference on Web Intelligence*. Retrieved November 21, 2008, from http://portal.acm.org/citation.cfm?id=1331834
Warner, M. (2002). *Publics and counterpublics*. New York: Zone.
Yakel, E. (2006). Inviting the user into the virtual archives. *OCLC Systems & Services: International digital library perspective*, 22(3), 159–163. Retrieved November 21, 2008, from http://www.emeraldinsight.com/Insight/viewContentItem.do?contentType=Article&contentId=1570008
Zimmermann, P. R. (1995) *Reel Families: A Social History of Amateur Film*. Bloomington and Indianapolis, IN: Indiana University Press.

CHAPTER THREE

Geo-linguistic Analysis of the World Wide Web

The Use of Cartograms and Network Analysis to Understand Linguistic Development in Wikipedia

THOMAS PETZOLD[1] & HANTENG LIAO[2]

This chapter discusses the usefulness of geo-linguistic analysis for Internet studies by presenting two techniques to frame and visualize the linguistic development of the World Wide Web, in particular the geo-linguistic development amongst different language versions of Wikipedia. An emergent research agenda has been set to explore the multilingual aspects of the Internet using, for example, a global perspective on Wikipedia research. And yet, there is a lack of theoretical and methodological tools for understanding the distribution and diffusion of linguistic materials online. The idea of geo-linguistic factors is introduced in this chapter to address these shortcomings and to respond to the study of a wide range of issues such as linguistic pluralism on the Internet or, more generally, the diffusion of innovation. Cartograms and network analysis are presented as two techniques that showcase the potential uses of geo-linguistic analysis. These two techniques of measurement and visualization indicate certain geographic and linguistic affiliations among languages. It is argued that although certain more developed language versions such as English and German may have central positions in connecting all languages, there exists another pattern that can best be explained by geo-linguistic factors. Finally, the limitations and implications of such findings and techniques are discussed, not only for research on Wikipedia but for Internet studies in general.

Introduction: Geo-linguistic Factors and Their Use

Media research about globalization and localization raises the questions about how media institutions and audiences are reorganized and realigned along or across boundaries. Meanwhile, the software that underpins computers, digital networks, and the Web has undergone a process of internationalization and localization that facilitates the adoption of related technologies for users of various languages and from different regions.[3] Leading global search engine Google, for example, provides different interfaces and tools for more than 170 local domains and 120 languages and variations such as Canadian English, U.K. English, and U.S. English.[4] Similarly, Wikipedia, the global user-created encyclopedia, has over 270 language versions, including improved flexible interfaces for variations of Chinese (Mainland simplified, Singapore/Malaysia simplified, Hong Kong traditional, and Taiwan orthodox) within the Chinese version of Wikipedia (Liao, 2009).[5] Thus, it could be expected that this process of internationalization and localization (which support the everyday interaction and usage in the digital environment) has also reorganized and realigned institutions and users in different ways. However, there is little research in this area that brings the seemingly "technical" issues into the supposedly "macro" media research about globalization and localization. Yet, their everyday impact on activities in the digital environment cannot be overlooked. Such neglect may be explained by the early monolingual (English only) development of the Internet. There is, however, little excuse for researchers now to ignore the fact that languages (along with regional factors) have increasingly become a significant aspect of the Internet. Therefore, while the factors of languages and regions are indeed important for media research in general and for Internet research in particular, how researchers can approach these with appropriate working concepts and cutting-edge techniques remains an open issue.

This chapter suggests that geo-linguistics (or geography of language), a small but emergent field that exists between socio-linguistics and human geography, may provide some concepts and techniques for Internet and media researchers to explore the political, cultural, and social implications of internationalization and localization in the digital environment. Geo-linguistics is concerned with the distribution of languages over time and space. Thus, geo-linguistic analysis can assist researchers and policy makers by critically investigating the "what, where, when, who, and why" (Cartwright, 2006) questions about languages at "international (macro), national (meso), and urban (micro) levels" (van der Merwe, 1993, p. 23). Indeed, some media scholars have taken similar paths in the past using the concept of "geolinguistic regions" (Albizu, 2007; Sinclair, 1996) to explore the role of languages and regions in areas such as international and national TV programming.

This chapter argues that geo-linguistic analysis can also inform our understanding of the linguistic development of the Internet, and associated temporal and spatial changes. Some new kind of geo-linguistic analysis designed for online linguistic development should be expected to generate useful insights for future research and policy making, and may provide tangible research findings to better understand the relationship amongst globalization, localization and the Internet. This chapter initiates research endeavors in this emerging field by exploring the findings of a preliminary study on the geographic and linguistic affiliations amongst Wikipedia language versions. The results are expected to be relevant for researchers to reconsider traditional issues such as identity politics, language politics, and media marketing (that addresses markets across various languages and regions) in the digital environment.

Geo-linguistic factors, be they implicit or explicit, have sometimes facilitated and sometimes hindered certain technology adoption. Users can both be empowered and conditioned by the geo-linguistic configuration offered in their respective digital environments. For example, Japanese-speaking travelers may have difficulties locating Internet cafés abroad that provide a Japanese-ready environment, that is, where they can read and type Japanese language correctly. Furthermore, users can find their search results more rapidly and efficiently if they can manipulate the geo-linguistic conditions of the digital environment. It might be a better strategy, for example, to find British government-related information by searching the British version of Google (google.co.uk) instead of using any other language version. Moreover, search activities depend on the users' capacity to type keywords in a certain language, which are submitted to certain search engines that may provide different services by considering the users' geographic and linguistic affiliations (e.g., Australian or Mainland China users who want to circumvent certain government-initiated restrictions pertaining to accessing particular websites). Therefore, Internet researchers must take geographic and linguistic factors seriously as they play an increasingly important role in software development as well as in everyday interaction in the digital environment.

Wikipedia as a Critical Observation Site of the World Wide Web

Wikipedia can be regarded as one of the most interesting sites of observation for issues pertaining to geo-linguistic analysis online. Firstly, it claims to be the "free encyclopedia" that anyone can write and edit. With its increasing number of language versions, Wikipedia is argued to be the most comprehensive website containing different linguistic materials in a single site. Therefore, it should be rich enough in linguistic materials for researchers to understand the geo-linguistic development of the global Wikipedia project as well as the larger World Wide Web. Secondly, because each linguistic version is governed and run by its editors, its language policy debate and development can be different from those traditional language-planning settings (Liao, 2009). Thirdly, because of the "inter-language links" that aggregate and connect all language versions of the same (sometimes only similar) entry, Wikipedia contains rich information about how and when certain languages are connected and linked, providing some indication about the development, diversity, and diffusion of the linguistic materials it aggregates and maintains.

Geo-linguistic analysis is useful to answer questions on the diffusion and diversification of languages within Wikipedia and, more generally, the World Wide Web. How Wikipedia handles linguistic diversity and how its exponential growth spreads geographically are but two pertinent questions that can be asked within a geo-linguistic framework. Wikipedia's geographic development enables us, for example, to understand its (lack of) popularity in different regions around the world that may be explained by (a combination of) economic, political, social, cultural, and (in particular) linguistic reasons.

The geographic development of Wikipedia has reached a point of saturation in some regions, most notably in parts of Europe and the United States. These are the areas where Wikipedia cannot expect further staggering growth in numbers but rather, in regional and linguistic diversity, stability and quality. This will affect some but not necessarily all of the original seven language versions (that reached more than 100 articles in the year of Wikipedia's foundation in 2001): English, German, Spanish, Polish, Portuguese, Dutch, and Swedish. Wikipedia's diversity depends on its diffusion into different regions and areas around the world. On its official statistics website (http://stats.wikimedia.org), Wikipedia currently lists more than 270 languages in total, and categorizes them into six of the world's regions plus one category of constructed language.[6] At the core of its geographic growth and diffusion, we can begin to discern Wikipedia's linguistic policy from the broader linguistic development.

A critical analysis of geographic and linguistic affiliations of any site of investigation on the Internet (and elsewhere) requires a thorough understanding on how the units of analysis come into existence. Thus, in order to understand geo-linguistic activity on Wikipedia we need to understand the linguistic policy process behind it, in other words how Wikipedia languages get approval or become rejected. A multiple-step process determines the addition or rejection of a new language version to Wikipedia. The application procedure is extensively documented and can be monitored at all times.[7] The specific requirements Wikipedia have for new language proposal to be approved are:

1. A new language edition must not already exist on any project of Wikimedia (of which Wikipedia is arguably the most popular one).
2. The language must have a valid ISO-639 1–3 code.[8]
3. The language must be sufficiently unique that it could not coexist on a more general wiki.[9]
4. A sufficient number of living native speakers form a viable community and audience.[10]

Wikipedia's linguistic policy has adapted a traceable language adoption procedure so as to involve the community and concerned individuals. This has allowed Wikipedia to grow and diversify and essentially to become the most linguistically diverse project in digital culture.[11] And yet, we may still ask how geo-linguistically distributed and diverse Wikipedia really is.

Geo-linguistic Analysis for Diversity and Diffusion Measurement

Wikipedia is not fully available to many people around the world. At the moment, it provides a platform for five percent of the world's 6,000+ languages. It comes as no surprise then that Wikimedia's top two emerging strategic priorities focus on expanding within large as well as midsized and under-connected populations (Wikimedia, 2009). While language versions of large populations (such as China and India) promise exponential growth in numbers, reach remains an issue (e.g., Wikipedia's Chinese competitors Baidu and Hudong). Even when diffusion (of Wikipedia) may be increased by its reach to Chinese and Indian language populations, the level of linguistic diversity does not necessarily increase accordingly.

For one thing, the rise of major Chinese and Indian language (Mandarin Chinese and Hindu) does not guarantee the online development of other dialects and minority languages that exist in China and India. Paolillo (2007) asserts that the "Internet could shift over to Chinese as the dominant language and not become any more linguistically diverse in the process" (p. 424). He points out that whilst Mandarin Chinese accounts for fifteen percent of the global population, simply adding Chinese languages could actually decrease linguistic diversity (ibid). It needs to be stressed that such empirical approaches (quantitative and qualitative) are rarely employed in debates on linguistic diversity and, even if they are, the process of information gathering is further complicated by inconsistent and outdated statistics (cf. Gerrand, 2007).

Linguistic diversity, however, must not be confined to the mere nominal listing or enumeration of languages. Such common practice of measuring linguistic diversity (counting the number of languages within a given sphere), should be considered only as the standing point. Further understanding is required to see its actual activities, where the participatory digital culture can provide some indication. A participatory understanding can most notably explain how access to and participation on the Web is relevant for a specific language and its speakers (and how it is not), and why languages with fewer speakers are potentially and actually disadvantaged against major languages (and why they are not). The participatory and networking potential of digital culture is generally expected to provide new data, practice, and ideas to advance the ideal of linguistic diversity. Still, it has been argued elsewhere that what constitutes sustainable linguistic plurality in digital culture is not the mere participation of the world's 6,000+ languages, but their mutual interaction—36 million language pairs is the ultimate extrapolation from current possibilities (Petzold, 2010). Researchers have to discern the current reality from the future potential. The authors thus believe that the "participatory understandings" as in the debates on digital divides and digital literacy must be complemented by understanding some realities about *digitally-enabled language interaction*. Geo-linguistic analysis provides techniques that enable us to critically investigate such activities. There is much to be gained from measuring the interaction and diffusion among languages online.

Two Techniques of Geo-linguistic Analysis

There exist many techniques which combine geographic and linguistic factors that can help researchers explore, connect, and understand geo-linguistic research. It is outside the scope of this chapter to provide a comprehensive review of these techniques, but two categories of such possibilities can be identified by the way geographic and linguistic factors are combined. One category is to make maps about languages by showing and analyzing the linguistic data on maps. Another category is to conduct network analysis about the relationships between languages by showing and analyzing their interconnections. Using Wikipedia's data, this chapter demonstrates a few cartograms and a network map, for each category respectively, in order to show the potential usefulness and pitfalls for further detailed analysis in the future.[12]

Choropleth maps and cartograms: Distribution, diversity, and diffusion

One conventional way to show linguistic data on a map is to make a choropleth map. In a choropleth map, areas are labeled, colored or patterned in correspondence to the statistical variable of interest. A straightforward way to show Wikipedia's linguistic development across the world is to draw a choropleth map to show certain indicators of development. For example, if the numbers of articles, editors, and so on can be used as a proxy for the development of certain language versions of Wikipedia, then a choropleth map can be useful to demonstrate the geographic distribution and possible affiliation of different Wikipedia projects. In Figure 1, a choropleth map of European national languages shows the number of Wikipedia articles of their associated national languages in graded colors, with English (en) as darkest color and Irish (ga) as brightest.

However, it might be controversial to map a certain language onto one or several regions since the idea of one region with one language may not be applicable. Indeed, the belief that one region, especially one nation, has one language is contested, and is related to issues such as linguistic nationalism (mobilizing linguistic resources for nationalist agendas) and multilingualism (more than just one major linguistic capacity in a community). For example, not all nation states have only one official language (e.g., Belarus has Belarusian and Russian; Ireland has Irish and English; and states like Belgium, Switzerland, and Cyprus have official languages that are official languages of other states). Also, some nation states share the same official language, such as German in Austria, Germany, Liechtenstein, and Switzerland. Due to the limitation of mapping along the boundaries of nation states, the creation of the following map adhered to certain "mapping rules" which results in each nation state only having one official language at present on the map. After all, it is as if we are assigning existing Wikipedia language versions to each nation state. Therefore, it should be noted that by creating maps for this paper we had to make a few decisions in order to connect official languages to nation states, a procedure which is obviously open to other arrangements and criticism. Firstly, it is straightforward to assign the region to the official language that no other nation state uses. Secondly, for those states which do have more than one official language, and if some of the official languages are already assigned (e.g., Russian for Belarus and English for Ireland), they are deliberately removed from that region. The results are that all European official language versions of Wikipedia are assigned, with Belgium, Switzerland, and Cyprus left out.

By considering the geographic factors in the choropleth map in Figure 1, it is relatively intuitive to see that English is the dominant language in the Wikipedia project while other western European languages seem to have substantial numbers of articles as well. The next tier is filled with Nordic languages and some official languages of Eastern Europe. In turn, the next layer has languages such as Estonian (et), Latvian (lv), Belarusian (be), Croatian (hr), Bosnian (bs), Albanian (sq), Macedonian (mk), Icelandic (is), Azeri (az), and Georgian (ka). Language versions that stand out on this map in terms of geographic affinity are possibly Russian (ru)

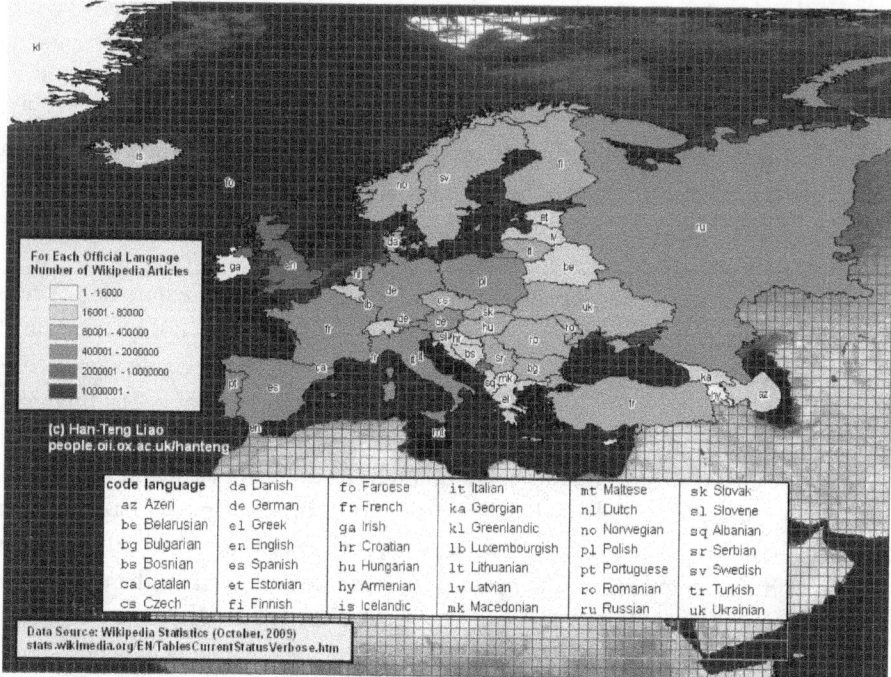

Figure 1. A choropleth map showing the number of Wikipedia articles for each national language in Europe

and Polish (pl), with more Wikipedia articles than the Nordic countries where Internet and information and communication technologies (ICT) development is expected to be more extensive. With this simple map, researchers can generate some ideas on how certain geographic and linguistic factors may influence the distribution or adoption of Wikipedia projects in Europe, and also consider the level of linguistic diversity inside the Wikipedia project for Europe.

Using the same dataset for a cartogram (Figure 2) shows a somewhat "distorted" map to readers. Indeed, the basic idea of cartograms is to substitute geographic properties such as area and distance with the variables of interest, so that the map appears distorted. Still, cartograms should not be regarded as deformed designs of maps, but rather as an alternative way to present a certain aspect of reality. As a graphical method it shows "the pattern of distribution of a single element" (Raisz 1938, p. 256) as opposed to a topographic map, which is made up of a combination of elements. For example, area cartograms are increasingly popular mapping techniques of showing the size of regions proportional to other different kinds of datasets, for example, GDP or population.[13] In other words, areas are adjusted to reflect the corresponding variables. As shown in Figure 2, due to the nature of cartograms, it becomes immediately obvious that English is dominant in Wikipedia's world of European official languages, with its large area representing a strong presence. Still, German, Polish, Dutch, and various Latin languages have strong showings considering their geographical size. In contrast, Russian has a relatively small size of Wikipedia articles in comparison with its geographical territory.

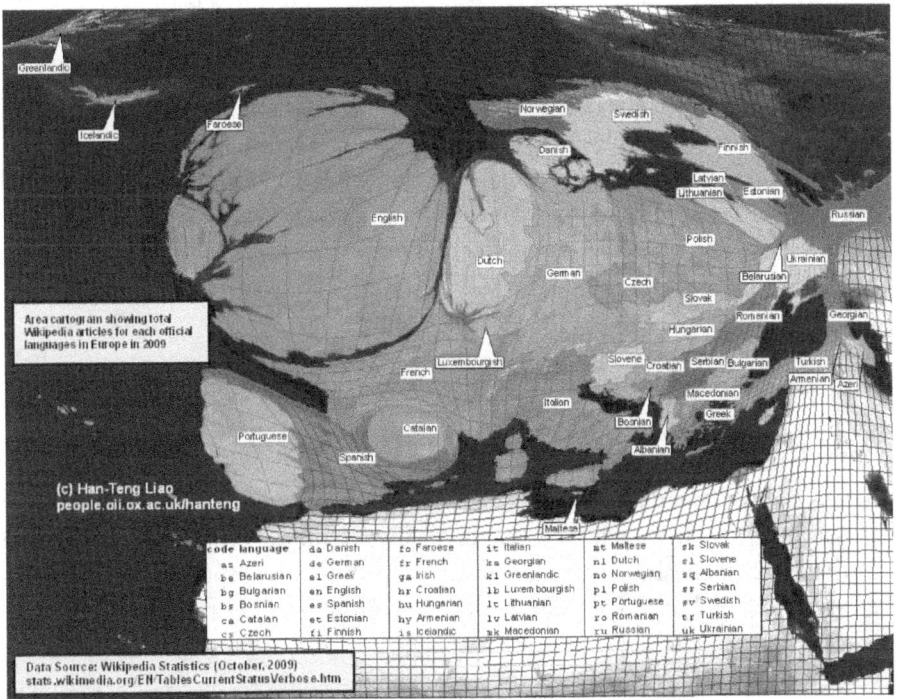

Figure 2. A cartogram showing the number of Wikipedia articles for each national language in Europe

By making a choropleth map and a cartogram with Wikipedia's dataset, researchers can begin to generate some preliminary observations and hypotheses, largely about the distribution and diversity of geographic and linguistic components of the global Wikipedia project in Europe. If similar map-making processes are repeated with Wikipedia's dataset for each year, the temporal elements can be included and examined. This chapter argues that such a technique is useful for studying the diffusion of innovation (within Wikipedia in this case) in order to consider whether geographic and/or linguistic affinity may have helped or hindered certain Wikipedia language versions to evolve. The geographic and/or linguistic affinity used in this paper refers to the connection and relationship that can be explained or shown by geographic and/or linguistic factors.

It is further suggested that such mapping of geo-linguistic factors of certain media development throughout a period of time can advance the concept and related discussion of "geolinguistic regions" (Albizu, 2007; Sinclair, 1996) for media research. It is particularly crucial to do so for Internet research because, unlike the areas of TV programming and film where national and regional boundaries can largely be assumed, websites such as Wikipedia and Google may not have clear segmentations when it comes to regions and languages. It remains an open question whether websites such as Wikipedia and Google have created certain "geo-linguistic regions" as in the research on global television. Still, with the techniques described above, it is feasible to collect certain datasets (web pages, users, activities, etc.) for a longitudinal study of the development along geo-linguistic lines. In this way, researchers may gain insights about not only "what" kinds of geo-linguistic regions have emerged but also "how" they have evolved.

This chapter particularly assumes that the use of cartograms can help researchers to understand the process of diffusion, an interdisciplinary concern categorized under the umbrella term of "diffusion research" (Bruce & Yearley, 2006). It is worth mentioning that the "diffusion research" in sociology, media studies, human geography, ethnography, etc. is borrowed from the concept of diffusion in physics. It is then understandable that computer programs, which make cartograms manageable, have been involved with the concepts of "distribution," "flows," and "diffusion." For example, Tobler (2004) describes the objective of cartograms within a physical analogy: "One may imagine that a thin sheet of rubber is covered with an uneven distribution of inked dots representing a distribution of interest. The objective is to "stretch the rubber as much as necessary until the dots are evenly distributed on the sheet" (p. 67). Moreover, "deliberate distortion occurs in order to make room for the symbols on an illustration depicting flow" (p. 58). The algorithm that created cartograms shown in this paper is designed by physicists who use a "diffusion-based" method to tackle computation problems (Gastner & Newman, 2004). It therefore seems appropriate to use area cartograms for diffusion research. For example, the cartograms of Figure 3 and Figure 4 indicate that the spread of Internet use in East Asia starts mainly from regions such as Japan, Korea, and the three Chinese-speaking regions of Hong Kong, Taiwan, and Singapore. Finally, it can be speculated that the growth of Internet users in mainland China may have been caused by its geographic affinity to all these regions plus a linguistic affinity with Hong Kong, Taiwan, and Singapore, especially with the large Internet population in the southern and coastal provinces of mainland China.

Network graph: affinity, core-peripheral, and diffusion

Another way to consider geo-linguistic factors is to draw a network graph, which aims to represent the interconnection between languages.[14] A preliminary network graph is presented in Figure 5, where each node represents a language version, with one-way and two-way directed links. The overall graph shows a core-peripheral pattern of the interconnections between different language versions of Wikipedia, with English (en), French (fr), German (de), Dutch (nl), Italian (it), Spanish (es), Russian (ru), Japanese (ja), and so on, constituting the core language

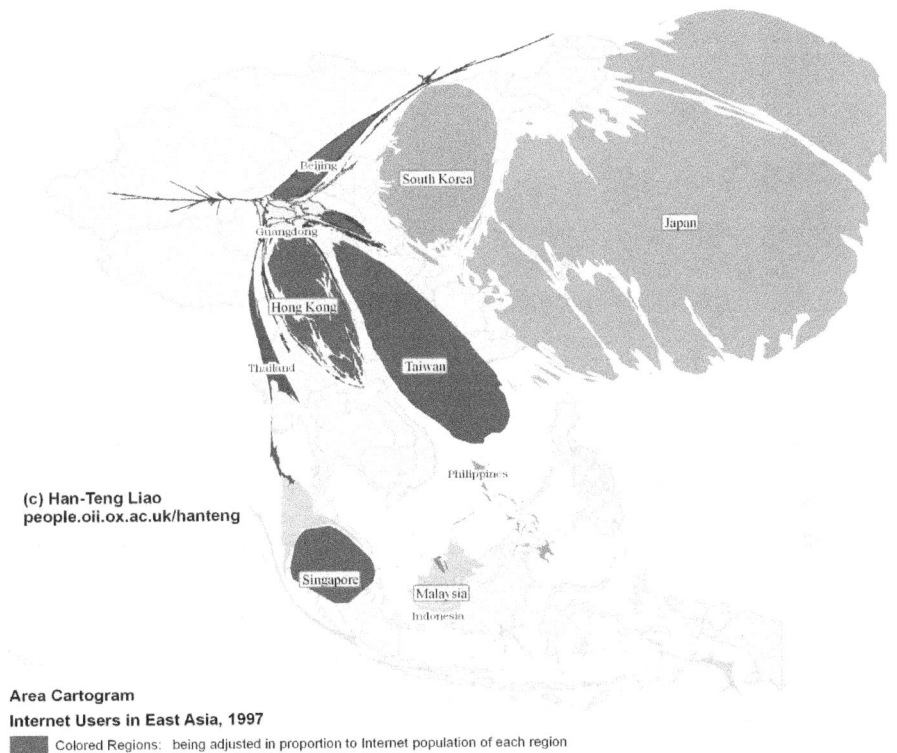

Figure 3. A cartogram of Internet users in 1997 East Asia

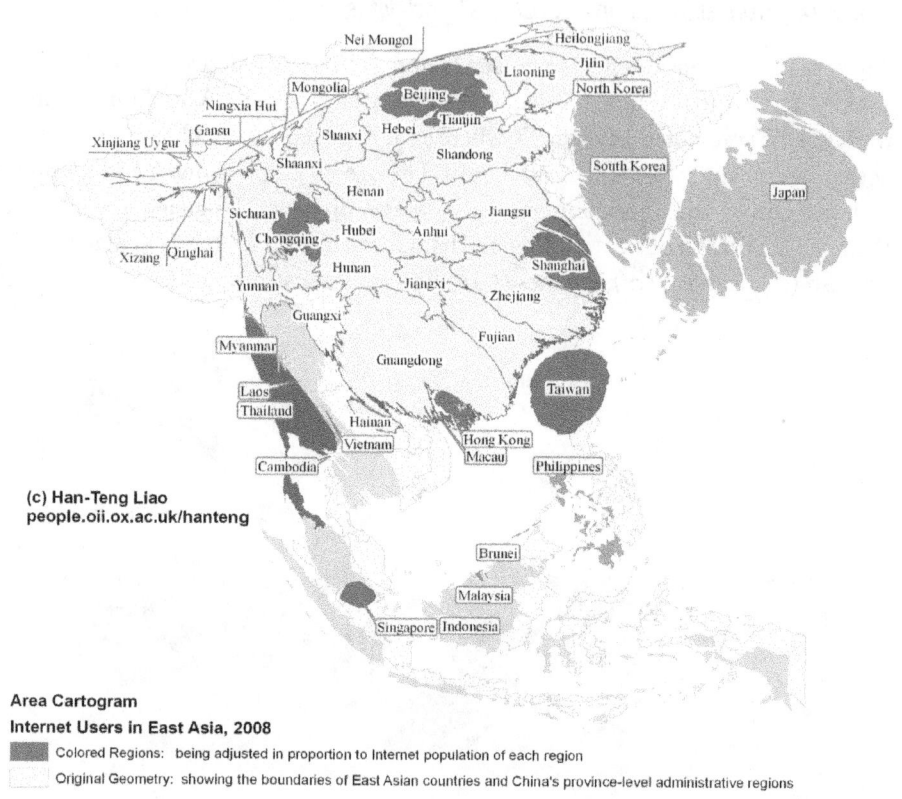

Figure 4. A cartogram of Internet users in 2008 East Asia

networks. Additionally, some observations can be made about geo-linguistic affinity shown in this graph. For example, Chinese (zh), Japanese (ja), Korean (ko), and Vietnamese (vi) are depicted as fairly close to one another on the top-center of the graph; while Spanish (es), Portuguese (pt), and Catalan (ca) are shown with similar affinity at the right. At the bottom center of the graph are the Russian (ru), Slovak (sl), and Polish (pl). This seems to reflect the geo-linguistic affinity of East Asian languages (such as Chinese, Japanese, Korean, and Vietnamese) as well as Latin languages and Slavic languages that constitute some kind of a group. In addition, the three groups are relatively kept away from one another, with other core languages such as English (en) and German (de) in the middle. By considering both the overall core-peripheral structure and the respective neighboring nodes, related hypotheses can be generated for further research, particularly with further data clustering tools.

The core nodes such as English and German may be moved away from geo-linguistic conditions because many other language versions of Wikipedia connect to them, no matter how far they are geographically or linguistically located from core languages. It can then be hypothesized that if a language version is connected by many other versions regardless of geo-linguistic affinity, the node that represents that language will likely be in the centre of the network graph, suggesting that they constitute hubs amongst many nodes (Barabasi, 2003), or in other words generate a sense of universality. For a language version that is only connected by few other versions, the node that represents that language will likely be situated around the nodes that are geographically and/or linguistically close to one another. In fact, the examples of Javanese (jv) and Sundanese (su), shown in the far right of Figure 5, suggest not only their overall peripheral positions in the global networks of Wikipedia, but also their geo-linguistic affinity with one another. If the network graph shown in Figure 5 truly represents the interconnection pattern amongst Wikipedia language versions, then it cannot be claimed that geo-linguistic factors are overcome by ICTs such as the Internet and Wikipedia. .

It cannot be claimed either that geo-linguistic affinity constrains every language. It is a much safer claim to make that geo-linguistic affinity is "distorted" somehow by ICTs in a way that the core languages may attain universal "connectedness," just as the languages of French, German, Russian, Japanese, and even Chinese are not only connected to the core node of English, but are also getting closer to one another for that reason. Moreover, it somehow reflects the infrastructure realities of global networks as well. These language versions are brought closer to one another not only because of English, but also because of the ICT development. For example, whereas Javanese and Sundanese seem to be left out on the periphery (on the far right-hand side of Figure 5) both of them have to connect to the network of languages via Indonesian (id), which is explained by their relationship to the Austronesian language family, and also reflects the hierarchy of international-national-regional languages, with Indonesian as a national language and Javanese and Sundanese as regional languages.

While the conjectures generated from the network graph shown in Figure 5 may be contestable, the purpose of the chapter is to show how geo-linguistic factors may or may not matter. In fact, the network graph of several language versions of Wikipedia has shown that geo-linguistic affinity may still matter, even without considering related geo-linguistic factors in making the graph. In other words, the geo-linguistic affinity is somehow reflected onto the network graph. The selected data used for generating Figure 5 is limited to those entries with less than three inter-language links, with the aim to highlight any affinity that can be captured by the spread of certain content in one language version to another. It should be noted that such affinity may be arbitrary for a specific entry. However, as the original dataset used for Figure 5 covers all inter-language links amongst all language versions of Wikipedia, such arbitrariness has been lowered by making sure there exists enough inter-language links for a link to appear in Figure 5. In other words, all the

68 | SECTION ONE: SOCIOCULTURAL INTERSECTIONS

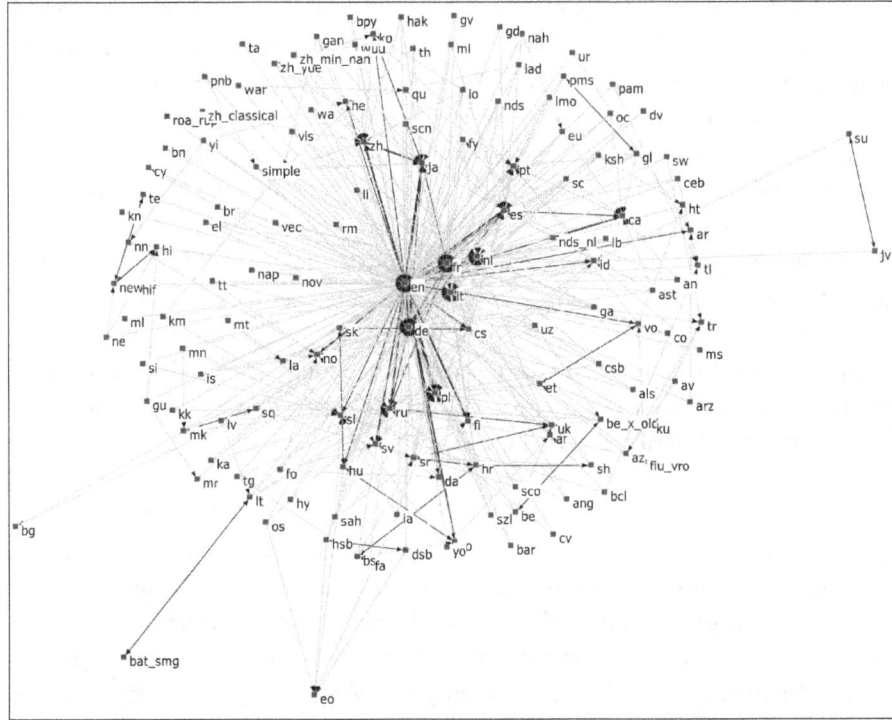

Figure 5. A network graph of selected language versions of Wikipedia

links that appear in Figure 5 suggest that there exists a substantial number of links between the two nodes of language versions, while if no link appears between two nodes in Figure 5 it means that there exist no or some unsubstantial number of links between them. Thus, it is important to clearly define the construct of links and nodes in order to create a valid and meaningful network diagram. Again, further vigorous network analysis is required to confirm the conjectures of core-peripheral structure and clustering effects (how and why some languages are more likely to then become neighbors than others), which is beyond the scope of this chapter. Nevertheless, the results so far do suggest that geo-linguistic affinity does exist and that the relationship is not arbitrary. This can better be explained by briefly describing how the dataset underpinning the network graph is actually collected, framed, and constructed.

The data on which the network graph is based is a selected set of the inter-language links amongst all the language versions of Wikipedia. To Wikipedia readers, the inter-language links show up in a box with the title "languages" at the left column of almost every Wikipedia entry article, which contains a set of links that lead readers to other language versions of the same (or nearly equivalent) entry, as shown in Figure 6, where only four other language versions (German, French, Japanese, and Chinese) exist for the English entry of the "Green Dam Youth Escort."

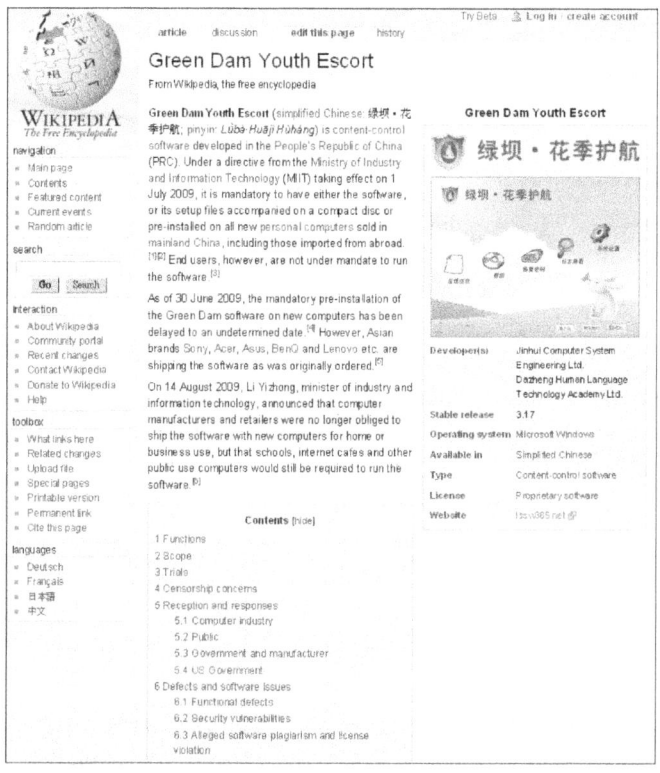

Figure 6. A typical Wikipedia entry page showing the location of the inter-language links

Such inter-language links can also be edited as part of the entry, with a straightforward syntax of the target language code followed by the title of the entry in that target language as shown in Figure 7.

Thus, from the example shown in Figure 6 and 7, the English entry of the "Green Dam Youth Escort" is linking to four other language versions. Most of the time, it is assumed for an entry to grow along with the increasing number of inter-language links. Certain popular (or universal) entries are destined to appear in almost all language versions, such as Wikipedia itself. Some regional or parochial entries are expected to be bound within certain language versions. It is worth mentioning that many new entries in Wikipedia start as translation of articles that already exist in other language versions, a process which the inter-language links help

```
[[Category:Science and technol
[[de:Green Dam Youth Escort]]
[[fr:Green Dam Youth Escort]]
[[ja:緑バ・花季護航]]
[[zh:綠壩・花季護航]]
```

Figure 7. An example of the wiki codes for a set of inter-language links

to facilitate. Hence, it is possible for some parochial entries to spread across other languages, and the number of inter-language links for a given entry can be thought to grow from zero (meaning the entry does not link to any other language version), all the way to the sum of all language versions minus one (meaning the entry has links to every other language version that exists). The growing dynamic nature of inter-language links thus provides an observation site where the patterns of spread, diffusion, and distribution of links can be shown by drawing the corresponding network graphs. At any given moment, it is expected that some entries may have more inter-language links than others, with possible values from zero to near the sum of all language versions. It is expected that those that have fewer inter-language links can help to cluster languages into meaningful groups whereas those that have more inter-language links can be regarded as having universally popular content. Hence, although the sample is limited in scope (e.g., the entry shown in Figure 6 is excluded in our sample), it nonetheless highlights important ties between language versions.

Also, similar to the discussion on using cartograms for diffusion research in the previous section, several network diagrams can be created at different times so that a process of diffusion may be observed. The use of network graphs for diffusion research is expected to have several extra benefits that may not be offered using area cartograms. Firstly, because the network graph may show a core-peripheral structure, suggesting some hierarchical relationships, researchers may observe how the spread of inter-language links reinforces, reconstitutes, or shifts the existing hierarchical relationship. For example, as languages such as Japanese, Chinese, Russian, German etc. grow in terms of number of entries, will these languages reinforce the current central position of English? Will the central node shift from English to another language? Or, will these languages reconstitute in order to become central nodes of similar significance?

Secondly, unlike area cartograms where geographic affinity is already assumed and presented on the map, the diffusion patterns observed from network graphs reflect actual linking affinity, which in the case of inter-language links of the

Wikipedia is likely to include geo-linguistic kinds of affinity. For example, it is conceivable that, borrowing from the concept of "geo-linguistic regions" (Albizu, 2007; Sinclair, 1996) for areas such as international and regional TV programming, some entries of popular culture may demonstrate how these cultural products may have been spread from one language domain to another. The dynamic dataset provided by the global Wikipedia project can thus be an important site of observation to empirically examine the geo-linguistic region concept, particularly as these inter-language links reflect actual developments online.

Thirdly, the use of network graphs can be regarded as an independent cross-check for the cartogram results because the geographic affinity is not assumed in the former while it is in the latter. For example, if some relationship that appears in the network graph cannot be explained by cartogram results, researchers will have to come up with explanations as to why such relationships may exist without clear geo-linguistic affinity. To sum up, conventional choropleth maps, cartograms, and network graphs can show the static distribution and temporal diffusion of languages using clearly defined constructs as well as reliable data. In addition, the network graph can not only show possible core-peripheral structures but also provide an independent cross-check of geo-linguistic analytics drawn from cartogram results.

Discussion

Some may argue that geographic and linguistic factors do not matter to central nodes such as English because of its dominant position in Wikipedia—every other language version connects to it somehow. Still, if we take a longer historical perspective on how the global Wikipedia project expands and diffuses from the core language version to others, one argument can be made that geo-linguistic factors are rendered hidden as English becomes "universal." The "universality" of English as a core language depends on the fact that English has been the principal working language of the Wikipedia project (and the Internet) and that the English version of Wikipedia has been used as a central point of contact for inter-language links. However, with increasing numbers of entries in other language versions, it is expected that some entries, for various geo-linguistic reasons to be explored, may only exist in a few language versions. These entries, as has been shown in this chapter, can provide some important indication about other geographic and linguistic development patterns that may reinforce or challenge the core position of English. Although no conclusive claims can be made so far (even within the case of Wikipedia), this chapter has shown why geo-linguistic factors must be considered, where and how to examine major geo-linguistic developments online, and what influence this has on issues such as linguistic diversity and innovation diffusion.

Our analysis serves as a showcase of more comprehensive research to come. For example, our analysis provides some preliminary explanations about how languages are linked or connected on Wikipedia. However, as to "why" this is the case, more research is required. The main contribution of this chapter is to show how the geo-linguistic factors may be important and how researchers can capture them. We have also showcased the usefulness of geo-linguistic analysis by using two techniques, cartograms and network graphs. Thus, the empirical evidence we use is mainly to show the possibilities and potentials of such techniques (and their underlying propositions), not to argue for or against an empirical argument on specific cases.

The importance of geo-linguistic analysis is crucial in understanding the development, diffusion, and distribution of human knowledge online and offline. After all, no matter how we define knowledge, it has to be written and debated in a certain language, a process which somehow conditions (if not determines) the development, diffusion, and distribution of ideas. The truism can best be observed in the case of the Wikipedia project, where some core language versions (e.g., English) can be identified as major vehicles for "universal" knowledge. However, it should be noted that it remains to be seen for alternative patterns or paths of development, diffusion, and distribution of knowledge to occur across different languages. This is why it is essential for researchers to frame, collect, and analyze the geo-linguistic factors that address issues such as linguistic diversity and diffusion of innovation.

LIMITATIONS AND FUTURE DEVELOPMENT OF GEO-LINGUISTIC ANALYSIS

Observing the World Wide Web with specific techniques and theories that take into account the potentials and limitations of geo-linguistic factors also require an awareness of potential limitations. Cartograms, for example, are used as a visual method for approaching a specific problem. They must not be misunderstood, however, as direct visualizations as they "can be hard to interpret without additional information" (Fotheringham, Brunsdon, & Charlton, 2000, p. 26). They are deliberately exaggerated and can be perceived as unusually displayed. The benefits of cartograms become most obvious when they are applied widely and in ways that a variety of agents (individuals, governments, enterprises) can relate to. Tobler (Id.) paraphrased this as follows: "Satellite image globes have to some extent supplemented political globes and anamorphic globes might someday also be constructed" (p. 69).

This chapter demonstrates only few possibilities of geo-linguistic analysis to show the benefits of analyzing how knowledge is developed, distributed, and diffused across languages. Two cutting-edge techniques (cartograms and network analysis) are deployed to explore the potentials of new ways of using geo-linguistic analysis. Nonetheless, the future expansion and refinement of such possibilities requires researchers to think thoroughly through the contexts and purposes of exercising geo-linguistic analysis online. The authors of this chapter strongly believe that research that uses and follows geo-linguistic data and analysis will contribute to various policy and research areas, for example, around the broader issues of information and communication technologies for development (ICT4D) as well as Internet governance. After all, they share the same concerns as to how certain geo-linguistic development patterns are made possible and rendered difficult. Finally, by showcasing two techniques applied to the global project of Wikipedia this chapter hopes to stimulate more methodical and conceptual experiments on using and rethinking this much neglected but essential component of Internet research and practice.[15]

NOTES

1. The author is grateful for the financial support towards this research that scholarships from the following organizations provided: German Academic Exchange Service (Bonn, Germany), Kurt-Tucholsky-Foundation (Hamburg, Germany), Creative Industries Faculty at Queensland University of Technology (Brisbane, Australia).
2. The author thanks Academia Sinica, the Oxford Internet Institute and PGP Corporation for the financial support (provided by the Academia Sinica Fellowship for Doctoral Candidates and the PGP Oxford Internet Institute Scholarship) and the technical support (provided by the GIS team at the Computer Center of Academia Sinica).
3. One particular indicator for such developments are numeronyms like i18n (internationalization) and L10n (localization).
4. For more updated numbers and a list of the local domains and languages see http://www.google.com/language_tools?hl=EN (last accessed February 15, 2010)
5. For more updated numbers and a list of the Wikipedias see http://meta.wikimedia.org/wiki/List_of_Wikipedias (last accessed February 15, 2010)
6. Wikimedia defines regions and region codes as follows: 'Regions are parts of the world where the language is spoken in substantial amounts (compared to total number of speakers). Regions where a language gained presence only by a recent diaspora are generally not included. Region codes: AF:Africa, AS:Asia, EU:Europe, NA:North America, OC:Oceania, SA:South America, W:World Wide, CL:Constructed Language' (cf. http://stats.wikimedia.org/EN/Sitemap.htm, last accessed February 15, 2010)
7. For Wikimedia's language proposal policy see http://meta.wikimedia.org/wiki/Meta:Language_proposal_policy (last accessed February 15, 2010)http://tiny.cc/B93yChttp://tiny.cc/B93yC
8. This means it must be listed in an ISO-639 database, or standards organizations must be convinced to create an ISO-639 code for a 'new' language.
9. This, in most cases, excludes regional dialects and different written forms of the same language.
10. This requirement, which must be met for the final approval, is discussed in an open discussion.

To do so, a project will be initiated where interest by individual speakers or supporters of the language is registered and arguments for and against the admission of the new language are gathered. Then a decision will be made by the language committee.

11. Cf. http://stats.wikimedia.org/wikimedia/animations/growth/index.html (last accessed February 15, 2010)
12. One useful tool for analyzing Wikipedia's statistics is the Wikimedia Toolserver which has provided necessary assistance for the relevant materials used in this chapter. Our particular thanks are due to Daniel Kinzler at Wikimedia Deutschland. Wikimedia Toolserver is hosted by Wikimedia Deutschland with the assistance of the Wikipedia Foundation, and provides sophisticated access to the Wikipedia database for researchers. For more information see https://wiki.toolserver.org/view/Main_Page (last accessed February 20, 2010)
13. Cf. http://www.worldmapper.org/ (last accessed February 15, 2010)
14. For a brief methodological note, the selected raw data was generated with the Wikimedia Toolserver, which was then processed with the programming scripts written by Han-Teng Liao to produce a network graph file. In a next step, the network graph file was fed into social network analysis and exploring tools such as NodeXL and UCINET. The tentative graph shown in this paper is produced by UCINET, with the spring embedding layout. The settings for the layout are as follows: the criteria is based on "Distances + Node Repulsion"; the starting positions "Gower scaling"; the number of iterations 30; the distance between components 30; the proximities "geodesic distances."
15. This book chapter was accepted for submission by the editors in late 2009, and a final version was submitted to the publisher in February 2010. In the meantime, other researchers enriched the emerging field that we outline in this study. Although we have not had the opportunity to discuss and include in this chapter works that were published after final submission, we were able to convince the publisher to include in our chapter what we believe is another important inquiry into inter-language links in Wikipedia: Hecht, B. & Gergle, D., The Tower of Babel Meets Web 2.0, CHI2010, April 10–15, Atlanta, Georgia.

References

Albizu, J. A. (2007). Geolinguistic regions and diasporas in the age of satellite television. *International Communication Gazette, 69*(3), 239–261. doi: 10.1177/1748048507076578

Barabasi, A. L. (2003). *Linked*. New York: Penguin.

Bruce, S., & Yearley, S. (2006). Diffusion of innovation. In *The Sage dictionary of sociology* (p. 73). London: Sage.

Cartwright, D. (2006). Geolinguistic analysis in language policy. In T. Ricento (Ed.), *An introduction to language policy* (pp. 194–209). Malden, MA: Wiley-Blackwell.

Fotheringham, A., Brunsdon, C., & Charlton, M. (2000). *Quantitative geography*. London: Sage.

Gastner, M. T., & Newman, M. E. J. (2004). Diffusion-based method for producing density-equalizing maps. *Proceedings of the National Academy of Sciences of the United States of America, 101*(20), 7499–7504. doi: 10.1073/pnas.0400280101.

Gerrand, P. (2007). Estimating linguistic diversity on the Internet: A taxonomy to avoid pitfalls and paradoxes. *Journal of Computer-Mediated Communication, 12*(4), 1298–1320, doi: 10.1111/j.1083-6101.2007.00374

Hecht, B. & Gergle, D. (2010), *The Tower of Babel Meets Web 2.0*, CHI2010, April 10–15, Atlanta, Georgia.

Liao, H. (2009). Conflict and consensus in the Chinese version of Wikipedia. *IEEE Technology and Society Magazine*. Retrieved March 30, 2009, from http://www.ieeessit.org/technology_and_society/default.asp

Paolillo, J.C. (2007). How much multilingualism on the Internet? Language diversity on the Internet. In B. Danet & S.C. Herring (Eds.). *The Multilingual Internet* (pp. 408–430). Oxford, UK: Oxford University Press.

Petzold, T. (2010). *36 Million Language Pairs: Generative Multilingualism in Digitally-Enabled Societies (Discussion Paper for the Twelfth Berlin Roundtables on Transnationality)*. Berlin: Social Science Research Centre. Retrieved March 10, 2010, from http://www.irmgard-coninx-stiftung.de/fileadmin/user_upload/pdf/Cultural_Pluralism/Language/Essay.Petzold.new.pdf

Raisz, E. (1938). *General cartography*. New York: McGraw-Hill.

Sinclair, J. (1996). Culture and trade: Some theoretical and practical considerations. In E. G. McAnany & K. T. Wilkinson (Eds.), *Mass media and free trade* (p. 444). Austin, TX: University of Texas Press.

Tobler, W. (2004). Thirty-five years of computer cartograms. *Annals of the Association of American Geographers, 94*(1), 58–73, doi: 10.1111/j.1467-8306.2004.09401004.x

Van der Merwe, I. (1993). A conceptual home for geolinguistics: Implications for language mapping in South Africa. In Y. J. D. Peeters & C. H. Williams (Eds.), *The cartographic representation of linguistic data (Discussion Papers in Geolinguistics, Nos. 19-21)* (pp. 21–33). Stoke-on-Trent, UK: Staffordshire University.

Wikimedia (2009). Emerging strategic priorities. Retrieved February 2, 2010 from http://strategy.wikimedia.org/wiki/Emerging_strategic_priorities

CHAPTER FOUR

Mapping the Future of News in a Digital World

US and Australian Perspectives

LUCY MORIESON & NIKKI USHER

News is changing in profound ways with the development of the Internet. Some have even gone so far as to call it the biggest revolution since the printing press—one we cannot fully appreciate because we are in its midst. As American journalism critic Clay Shirky (2009) has maintained, "when someone demands to know how we are going to replace newspapers, they are really demanding to be told that we are not living through a revolution." This chapter takes these revolutionary changes as its starting point to examine the print media of the United States and Australia as they face the challenges found in the nexus of digital technology and the institution of news. We ask questions about how the Internet is challenging the dominance of the traditional news model and its economic, political, and cultural position in society, and we look at what the loss of these legacy models might mean for journalism as an institution. By exploring the differences and similarities between the state of the two countries' print media, we hope to suggest points of future discussion for studies that explore the influence of the Internet on the transformation taking shape in news. First, we begin by looking at some of the broad changes facing journalism, and then move to particular case studies: *The New York Times* in the United States and Fairfax Media in Australia. While we work to compare the similarities across our cases, we also explore the differences between the two countries' media landscapes and histories and discuss the role and significance of print news in countries with different media ecologies; the United States, without a strong public media presence, and Australia, a country with a strong public media but with a diminishing investment in print news. Working through these case studies, we hope to glean insights about what parts of journalism we ought to salvage from the past, and what we can learn about journalism from its reinvention online.

CHANGES, CHALLENGES, AND OPPORTUNITIES FACING JOURNALISM IN THE DIGITAL ERA

Journalism is undergoing a fundamental re-evaluation of its underlying core principles and ideas. Notions of who can speak as a journalist, what is considered journalism, journalism's role in a democracy, the role of the audience, and the very basis on which journalism has operated as a viable business are all being re-evaluated as the institution of journalism intersects with digital technology. In both the American and Australian contexts, the professional status of journalism is being questioned and challenged as journalists grapple with their position within a changing professional culture and an increasingly commercialized media infrastructure. As Carey (1969, 1996) and Schudson (2003) discuss, professionalism for journalism has long been associated with education, the development of professional codes of ethics, and affiliation to news organizations. All of this has been challenged by the rise of the Web and the capacity for individuals to produce news that can be distributed to mass numbers of people, regardless of professional status or affiliation. Through blogging, social media tools, and the rise of forms of citizen journalism, the tools of news production are increasingly in the hands of what Rosen (2006) has described as "the people formerly known as the audience." Despite these challenges, journalism as an institution continues to cling to its professional norms and practices—from ethical codes and associations, to workplace structures, temporal news structures, and relationships with audiences.

As the focus on participation and nonprofessional production practices suggests, the role of the user becomes central to journalism in the digital age. Benkler (2006) has reflected upon the changed architecture of the news information ecology in the Internet age, arguing that the hub and spoke model of mass media, where news was collected by a few and distributed to many, has been replaced by a networked public sphere of collaborative information gatherers and distributors. In this environment, even those on the outer nodes of the network have the chance to be heard. If the adage was that "he who owned the press could print the news," then access to printing has been significantly democratised, as everyone with a computer has potential access to the means for crafting news. Furthermore, new generations of news audiences in the United States will be completely unlike past generations, having grown up with the Internet, and not likely to turn to print news later in life as previous generations have done (Brown, 2005). Beyond issues of readership, the physical consumption of news has changed, becoming a task that can occur in any place and at any time using a digital device, no longer facilitating what Benedict Anderson has called that "extraordinary mass ceremony: the almost precisely simultaneous consumption...of the newspaper as fiction" by a national audience of readers (1991, p. 35). Now, news punctuates desk workers' work cycles, and the news that is found online is far more closely aligned to the hourly cycle of the broadcast news cycle than the daily cycle of print news.

Given the challenges already outlined to news production, distribution, and use, it is no surprise that one of the greatest issues facing news businesses in the digital era is adapting their business model to these changing conditions. In this regard, the main challenge faced by print newspapers is attracting advertisers to a product that is no longer as appealing to both businesses and audiences as it was during print's prominence (Beecher, 2009a). For online versions of publications that originated in print, the challenge is how to coexist alongside their print versions, with possible visions of ultimately replacing them, but as yet, existing on an unsustainable revenue model. For online-only publications, the challenge may be similar—how to monetize what has traditionally been a free product—or it may be the case that they are one of the lucky few outlets that have managed to find a niche market that is willing to pay for online content given its particular value to the audience—The *Wall Street Journal* and Australian publication *Crikey* are two such cases. In either scenario lies the challenge of trying to adapt an existing product to changing business conditions as audiences and authority become dispersed online.

Along with these challenges to journalism's professional status come questions about journalism's role in a democracy. The term "citizen journalism" begins to suggest some of the optimism that has surrounded the development of new forms of communication online, and much has been made of the democratic potentials of the nexus of Internet technologies and news (Gillmor, 2004; Glaser, 2006; Lasica, 2007; Rosen, 2006). While it is important to recognise the persistence of such rhetoric, it is also necessary to note that such optimism is beginning to be replaced by a more cautious tone. Simply because more people are involved in creating journalism does not mean that we have a more democratic society, argues Schudson (2008). In fact, it is important to recognize that traditional forms of journalism have played a significant role as a fourth estate check to power holders, precisely because these news organizations were themselves powerful institutions. Thus, a tension is played out between the opportunities for engagement as ordinary people create news, and the decreasing vitality of professional journalists at news organizations and their ability to question authority. While it is appropriate to be optimistic about the former, here we are concerned with the latter, and how new incarnations of news might preserve some of journalism's democratic role.

Despite this range of challenges, we argue that there is still much need, value, and demand for quality, serious, public-oriented journalism. Despite the excitement surrounding the democratic potentials of digital communication, publishing, and distribution models, there is a vital need to preserve the public service function enshrined in print news, and to ensure that this is translated into new spaces online. The public service model of the news understands journalism's role as a "service in the public interest, one that is shaped with an eye towards the needs of a healthy citizenship" (Zelizer, 2005, p. 73). As Kovach and Rosensteil (2007) note, journalism's first responsibility is to its citizens. The public service orientation can also be considered the "watchdog" function in democracy, holding public officials and others with power in check. As Curran (2005) explains, this public service orientation provides representation, enabling groups that wouldn't otherwise be given the chance to be heard. At its most basic, the news tells us things we do not know (Schudson 2008), providing an important conduit between institutions and an informed public. In these ways, news continues to fulfil an essential role in providing information, a forum for discussion and deliberation, and as a check for institutional power.

One of the most important things about newspapers is the journalism they provide. Nichols (2007) argues that newspapers have the reporting infrastructure for in-depth coverage, are credible to the people they are covering, and form the broad strokes of the public agenda both locally and nationally. Downie and Kaiser (2003) add that newspapers do the most original reporting, compared to other forms of news media. Starr (2009) argues that newspapers have been the "civic alarm systems" by keeping their eyes on the state, covering a much broader range of stories than broadcast. Newspapers, according to Starr, are the heart of public affairs coverage and are responsible for rigorous fact-checking and scrutiny. This role remains important despite blogs establishing themselves as a central part of the news cycle. Kelly (2008) found that even the most extreme sides of the political blogosphere rely on mainstream news reporting to make arguments, highlighting the necessity of news organisations' investment in paid reporters and investigative journalism. In light of these changes to the production, distribution, and use of journalism, let us now consider the details of our case studies. We will look closely at the way in which the news industries of these two countries are experiencing and responding to these challenges.

THE STATE OF NEWS IN THE UNITED STATES

The United States is facing a dismal future for traditional news. The number of younger people getting no news on an average day is increasing; 34 percent of

Americans between 18–24 report getting no news on an average day, up from 25 percent in 1998 (Rainie, 2009). Print newspapers in the United States have faced a slow and steady decline over the past two decades, averaging approximately 1–2 percent per year ("State of the News Media," 2007; "A Graphic History of Newspaper Declines," 2009). In 2008 the decline was 4 percent, and from 2001 to 2008, the decrease for daily circulation totaled about 13.5 percent (State of the News Media, 2009). In some markets, circulation at the largest newspapers dropped tremendously in 2009. The *San Francisco Chronicle* lost 25 percent of its readers in a single year; the *Houston Chronicle* lost 14 percent of its readers (Stark, 2009). Two major US newspapers closed their doors in 2009: the *Rocky Mountain News* and the *Seattle Post Intelligencer*. Both of these newspapers were in two-newspaper towns, leaving even fewer cities with competing metropolitan dailies. The financial forecasting and auditing company Fitch Ratings predicted that there would be at least one major metropolitan city in the United States without a newspaper by the end of the decade ("Coming Soon? The No-Newspaper City?," 2008). All of the bad news seems ironic, though, when Americans are turning to online news now more than ever before; Pew reports an estimated 1856 percent jump in online news readership since 1996 (Rainie, 2009); 29 percent of people report reading online news up from 23 percent 2 years ago (State of the News Report, 2009).

The recession of 2008 only accelerated what was already a rising problem in the newspaper revenue model—the loss of classified advertising to the Internet. The advent of Web sites like craigslist.com, which advertises jobs, apartments, and personals for free (or close to it), as well as the decline in department store advertising and the near-death of the American car industry have all contributed in the drying up of what were once newspapers' "rivers of gold"—the classified pages. Yet newspapers make 90 percent of their profits from print (Hirschorn, 2009), so to cut off this organ of profit is a terrible idea. As Hirschorn points out, the 1 million *New York Times* subscribers are far more valuable to advertisers than the 20 million unique Web users each month. He claims that, "common estimates suggest that a Web-driven product could support only 20 percent of the current staff" at the *New York Times* (ibid). Starr (2009) echoed a similar sentiment when he expressed distaste for the decision that *The Detroit News* and the *Detroit Free Press* made to cut home delivery to 3 days a week. Not only is the paper making a short-sighted decision by cutting off its biggest revenue stream, but it is also stemming people's habit for the print paper and pushing more readers online—where they "may find alternatives to local papers and never come back" (Starr, 2009).

The Case of The New York Times

The New York Times is America's newspaper of record and the winner of 101

Pulitzer Prizes. Its online site is consistently America's highest rated newspaper Web site, and it often breaks into the top five of all online news sites ("Top 15 Most Popular News Web sites," 2010). Despite being one of the best newspapers in the world, *The New York Times* is suffering from tremendous financial difficulty, in part because it has failed to adequately find ways to monetize its audience. Providing top-level reporting is costly—the Baghdad bureau alone costs $3 million a year to fund (Mnookin, 2008). So what kind of situation does America's paper of record find itself in? And what does this say about the newspaper industry in America overall? *The Times* has progressed to a place—at least in management's perspective—that most newspapers are striving to reach: where print and online no longer compete with each other but instead are fully integrated. Editor Bill Keller felt so confident about this that he saw no need to replace the integration czar at the *Times* when he moved to another section of the paper, noting "there might be no equivalent to that role in the future" (Perez-Pena, 2009a). This suggests that many of the old wars between print and online are solvable.

But financial problems remain. *The Times* is far from economically stable—for instance, there was widespread rumormongering among industry insiders that *The Times* might collapse in Spring 2009 because it was due to deliver $400 million to lenders. The then $1 billion debt was not just the result of a poor economy—the newspaper company had overextended itself by purchasing its new headquarters for $600 million (Hirschorn, 2009). With the looming debt, *The Times* took cost-cutting seriously. Several weekly sections were cut to save money, while newspaper employees took a five percent pay cut to avoid layoffs in 2009 (Arango, 2009). *The Times* seemed more stable by Summer 2009, turning a profit of $US 39.1 million; further, the bad news of an ad revenue drop was offset by the paper's cost-cutting measures, according to Perez-Pena (2009b). But financial troubles remain. The company cut its staff by eight percent at the end of 2009, eliminating 100 newsroom jobs by layoff or buyouts—"the first time in memory that had happened" (Perez-Pena, 2009c). Many point to *The Times'* experiment with "TimesSelect" as evidence of a failure for specialized content. *The Times* did gross $US 10 million from the experiment and was able to charge higher advertising rates, but also realized that people were coming into the Web site through search engines—and they were losing all of this traffic. *The Times* is rumoured, according to a number of industry blogs and insiders, to begin introducing a new subscription model during 2010—for all of its content—as a way to build revenue ("Times ponders online fee$," 2009).

Why does the United States need *The New York Times?* For one, it serves an agenda-setting function for public discourse because it has the resources, readership, and institutional recognition to create dialogue about issues of public concern. Its institutional heft gives it professional journalists who are trained in source cultivation and development, careful investigation, and the legal backing to keep from getting sued when it threatens to expose wrongdoing. As an institution it plays a formidable role in creating and maintaining democracy by providing systematic, daily (and even up to the minute) online, print, and multimedia coverage of major issues and institutions in the United States and worldwide. At the same time, *The New York Times* is only as strong as the audience it engages. The vibrant participatory media of Web 2.0 needs to be part of legacy media in order for it to remain responsive to the interests and needs of the community it serves. As *The Times* refines its digital strategy, it should look to find ways to incorporate the audience as content creators and providers of wisdom, rather than being the sole director of the conversation. *The Times* has begun to encourage conversation with comments, Twitter, and blogging, but is yet to reposition itself as a forum for discussion, rather than a speaker in a one-way conversation with its readers. As such, we can learn from the ways that journalism has reinvented itself online and use these transformations within a "legacy" news setting.

For all of its problems, *The Times* embodies the tensions of a changing business and technological environment. The future of the newspaper is carved out by immediate survival tactics, but a new revenue model to replace the old is still the missing link to success. Layoffs and buyouts are being used as ways to cut costs, but these efforts ultimately undermine the quality of news produced, hurting the core product. At the same time, there are record numbers coming to the Web site, but the news organization has no idea how to successfully monetize these hits, and online advertising has failed to provide a boost. Whether pay sites will work remains to be seen, but walled gardens on the Internet are often more porous than they appear. In the meantime, the lack of a new business model for American newspapers places the continued survival of massive news-gathering institutions at risk. Our discussion will suggest some ways for news organizations to rethink their current models.

THE STATE OF NEWS IN AUSTRALIA

While the trends in declining engagement with traditional media in the United States seem to be echoing across the globe, Australia has a long way to go before matching the decline seen there. But, as in the United States, newspaper circulation in Australia is in a long-term decline. Readership of weekday editions of print newspapers fell by 21 percent between 1993 and 2005, but this can not be conclusively linked to Internet use (Este et al., 2008, p. 9). On closer examination, more recent figures are not as troubling as much of the debate and hyperbole might indicate: between 2006 and 2008, circulation of major Australian daily print newspapers dropped only 0.7 percent ("State of the News Print Media in Australia,"

2008). But as in the United States, online news figures continue to indicate that increasing numbers of Australians are turning to the Web to source their news. While news producers have yet to find ways to make a profit out of this activity, audiences have demonstrated their willingness to use the Web as a news source, particularly in Australia, which has one of the highest percentages of online news visitors in the world ("Life in the clickstream: the future of journalism," 2008, p. 11).

Reading the news is the third most common activity undertaken by Australians on the Internet, after email and banking (*ACMA Communications Report* 2007–2008, p. 51), and news portals like ninemsn and online newspapers like smh.com.au consistently appear in the nation's top 20 Web sites (*ACMA Communications Report* 2007–2008, p. 53; HitWise, 2010). Despite this relative success, Australian media commentators and professionals have been preoccupied in the past year by what is seen as the decline of sources of serious journalism in the Australian media. The response from Australia's independent media has been to lobby for a source of publicly funded journalism that will support serious, high quality, public-oriented journalism for the good of the nation's democracy (Dikeos, 2009; Sparrow, 2009). So, while digital media is providing a platform for much of this debate, concerns remain about both the position of print newspapers and the status of journalism in relation to the democratic process.

The Case of Fairfax Media

Australia's Fairfax Media company has traditionally been positioned as the main competitor and left-leaning political alternative to the Murdoch family's national dominance. Fairfax runs the two most successful broadsheets in the country—*The Sydney Morning Herald* (*SMH*) and *The Age* in Melbourne—as well as the nation's only daily business tabloid, *The Australian Financial Review*, among a handful of other publications in Australia, New Zealand, and the United States. Since a 2007 merger with Rural Press, the company has repositioned itself as an "integrated media company" (Fairfax Media, 2009)—consolidating its operations across various media into one umbrella company. This was followed by moves to expand their digital network, Fairfax Digital, which was positioned as a portal for access to the company's more than 30 news and classified Web sites. It also saw the company's expansion into online-only news, first with the 2007 launch of brisbanetimes.com.au, then in 2008 with the launch of watoday.com.au—each move significant for the decision to launch a digital product to compete with the print product in Perth and Brisbane, which are one-paper cities.

Fairfax has always occupied a significant position in the Australian media landscape, due to its status as the key competitor to the Murdoch-owned News

Corporation papers. This position is important firstly because it provides an alternative political perspective to the predominantly right leaning and conservative News Corporation papers, and secondly because *The Age* and the *SMH* have always been seen as the home of "quality" journalism in the country. It is important to grasp both elements of this significance. Where Fairfax was not competing with tabloids with little interest in "serious" journalism, they were competing with *The Australian*, News Corporation's Australia-wide broadsheet, which in recent years "had consciously positioned itself as the site for thought leadership in conservative politics in Australia" (Flew, 2008, p. 9). The paper vigorously supported then Prime Minister John Howard through a range of contentious political moves, including joining Bush's war on terror—a time at which the mere existence of the Fairfax papers was seen as an essential alternative in the national media landscape. During this and other key times in contemporary Australian political history, the Fairfax papers have been important not only in their ability to offer an opposing political stance, but simply in their scope to provide a forum for serious, reasoned discussion and deliberation.

But recent financial difficulties have threatened the company's capacity to offer this forum for serious journalism. The most recent and dramatic indication of these difficulties occurred in August 2008, when the company announced plans to cut 550 members of its staff across its various branches and operations in both Australia and New Zealand (Zappone, 2008). This news was particularly poorly received as the company had the week prior posted a profit of $AUD 386 million, which was up 47 percent on the previous year (Ricketson, 2008). The move was interpreted locally as signalling that Fairfax was no longer interested in providing public-oriented journalism in Australia (*Trends in Newsrooms: The Annual Report of the World Editors Forum*, 2008). Since these events, the concern over the place for "quality" journalism in Australia and who will fund it has occupied the media industry and its champions and commentators. Meanwhile, Fairfax have named former CEO of the Woolworths grocery chain Roger Corbett as their chairman, eliciting concerns about his qualification for the task and the direction he will take the company in, when leadership is needed most (Beecher, 2009b).

In Australia as in the United States, the business model is central to these discussions about the future of news. But in the Fairfax example we can see the more pressing concern is that of the newspaper's role in a democracy. Given the peculiarities of Australia's media landscape, the existence of a competitor to the Murdoch press is vital to ensure a plurality of views. While Fairfax has been successful in maintaining a reasonable online presence, they have done little to facilitate new forms of audience participation online. Nonetheless, in creating two new Web-only news sites, they have signalled that they see the investment in online news as a serious long-term venture. Perhaps more interesting is the recent bundling of their commentary across all titles online under a relaunched *National Times* masthead (it formerly operated as a print weekly). Alongside independent sites like *Online Opinion*, the *National Times* is indicative of the growing popularity of opinion online—both for audiences, for whom it stimulates discussion and participation in online communities, as well as for news producers, for whom it is cheap and easy to make. While this sort of engagement should be fostered, the question remains: who will continue to invest in serious investigative journalism in Australia?

COMPARING THE US AND AUSTRALIAN CASES

It is important to note that we write this at a time of momentous change, and as such, are marking only temporary battles in the larger struggle to understand the place of legacy print news organizations in the new media world. Nonetheless, the struggles of *The New York Times* in the United States and the Fairfax papers in Australia occupy a similar fundamental premise—the advertiser-supported model of print journalism no longer works in today's Internet era. The two case studies underscore the difficulties faced in two Western countries with different media systems and imperatives—the United States, without a strong backbone for public media, and Australia, a country with a strong history of public media. Nonetheless, newspapers remain private media subject to the whims of capitalist fortunes—for better or for worse. But their fortunes in each country are significant, for newspapers continue to provide much of the original content for other news outlets, and serve an agenda-setting function for the rest of the news media and the public (McCombs, 2004).

While Australia has a long history of public broadcasting, the 1983 change of the name of the Australian Broadcasting *Commission* (ABC) to the Australian Broadcasting *Corporation* signalled deeper changes. The move towards internal corporatization put the broadcaster under increased financial pressure, threatening its ability to provide the in-depth investigative journalism for which it had become known. In this media environment, the existence of a media company to challenge Murdoch's monopoly on newspapers has been essential to Australia's media plurality—and ultimately, the country's democracy. Fairfax has traditionally fulfilled this role, but with the success of its print operations under threat, it is unclear how long it will be able to continue to do so. While its online ventures have proved successful, they have so far been unable to generate the sort of revenue that funded the newsroom traditionally required to support the sort of high-quality, long-term investigative reporting that secured Fairfax its important place in the Australian media and democratic landscape. The question of who, if anyone, will provide this sort of journalism as we move towards an increasingly digital future remains the key challenge facing the Australian news media.

In comparison, the United States does not have the same history of public broadcasting, though National Public Radio boasts approximately 32.7 million listeners a week ("NPR Reaches New Audience High," 2009) (however, it only receives about 2 percent of its funding from government sources ["Annual Reports," 2010]). Instead, the American public has mainly relied upon newspapers as the primary source of original news reporting in their communities. Independent news organizations have provided an objective source of news coverage that strives to keep news apart from bias and political opinion. With a diminished future for local and national newspapers, our concern is that there are few promising alternatives that have the scale, institutional heft, and audience to keep Americans aware of local, national, and international issues. Non-profit online journalism is one option, but it currently operates more like a weekly newsmagazine than as a reliable supplement to daily news reporting (Downie and Schudson, 2009). Our concern is that national issues will shift to increasingly polarized cable networks such as Fox News (and increasingly MSNBC) that do little more than provide opinion, creating an environment of bitter, one-sided, and conflict-ridden partisan journalism that limits national discussion and devotes little attention to local issues.

By comparing the scenarios facing media organizations in the United States and Australia, we can draw a number of broader conclusions about the state of the news media in each country. First, there remains in both countries a substantial appetite for news. However, the way in which audiences go about sating this appetite is changing; primarily, audiences are becoming increasingly fragmented. There is also a greater demand for opinion online, a form which stimulates audience participation and interactivity by breaking down barriers between the producer and consumer. However, this trend comes with its own concerns, as we have indicated. Secondly, in the digital environment, there is a need for news organizations, particularly those with a print provenance, to respond to audience demands for new forms of interaction and engagement. This does not have to compromise the style of news that has been the traditional preserve of a media company, but it is necessary for news companies to make spaces available to facilitate participation. Third, we believe that print media companies, even if they are no longer operating in print, must build on what has traditionally been their strength: the provision of high-quality journalism, and the necessary resources and journalists to undertake it. Fourth and finally, we can conclude that if this is to happen, then media organizations need to think creatively to address the specific business challenges that they are faced with to fund the sorts of ventures they are involved in. This requires providing a response (or a range of responses) to the problem of the "business model" that continues to dominate discussions about digital news. We do not believe that this will mean finding a business model "template" which will be universally applied, but rather, thinking creatively and experimenting with new ways to fund news in print and online.

WHAT LIES AHEAD?

As we noted earlier, tremendous opportunities exist for journalism in the digital era, and while we are interested in arguing for the preservation of some of print's qualities, it is equally important to consider what new forms of journalism can provide. American newspaper critic Jeff Jarvis argues routinely in his blog that newspapers are "stale," that they are not "conversations," that they are about "control," and that they are stuck in a paper mindset (Jarvis, 2009). Shirky (2009) has also proclaimed that no one really needs newspapers anymore—but that we do need journalism. Lyons (2009) similarly argues that newspapers should "die and get out of the way." These arguments rely on broad faith in the new initiatives offered by the Web, particularly when it comes to citizen journalism and social media. New opportunities for news creation abound in ways that both seek to provide alternatives and complements to mainstream news. These alternatives and additives include social media networks, a mix of specialized non-profit and for-profit Internet news sites dedicated to local, investigative, and foreign reporting; NGOs; and, of course, blogs, among others. Many of these involve the public in new ways, reconfiguring relations between creators, content, audiences, and notions of the public.

Recent years have seen a concerted emerging international effort dedicated to saving the institution of public journalism and the news more generally. There have been countless conferences dedicated to finding a new business model to support news. Part of this business model needs to include a reconfiguration of the audience in a way that sees the audience not as disconnected from the process of gathering news, but as a vital part of the news production process. By making individuals feel connected to the acts of creating and gathering news through an institutional arrangement, we can combine both the strength of crowd-sourced and conversation-based journalism with the heft of institutional legacy journalism. A good business model will find a way to capitalize on enhancing the many new voices that are part of the conversation—and keep them talking.

Among the possible solutions put forward, the three most prominent are the non-profit model, the public model, and the micropayment model. While some newspapers, such as *The Guardian* (UK), are already working successfully off the non-profit model, others have argued that newspapers are a $US 35 billion a year industry—and that endowments and philanthropists simply can't try to compensate for that (Westphal, 2009). In the United States, as in Australia, there have been arguments about the value of increasing government support of the industry, which would see newspapers receive funding based on similar models to public broadcasting (Pickard, Stearns, & Aaron, 2009; McChesney & Nichols, 2009). But for many news organizations, regardless of journalism's public or democratic role, the more pressing question is how to monetize the audience for news. One of the most prominent suggestions is micropayments—which would work somewhat like the iTunes purchasing model, whereby individuals would pay to access individual articles. Other systems would create subscription models much like subscriptions to print models, whereby subscribers could see premium content, but nonsubscribers could only see snippets. Those in favor of a user-pays system argue that information should be paid for and that newspapers offer specialized content that can't be found anywhere else. Rupert Murdoch agrees—he recently announced plans to introduce paywalls to his major news sites over coming years (Usher, 2009). But those who oppose paywalls say Internet loopholes and the nature of the Web make them redundant (and hurt Web traffic) (Salmon, 2009). One suggestion we make to keep news institutions solvent is to create a tiered system of content, whereby users have access to a main site but are given exclusive content if they pay. Such tiered systems have shown promise in online content as diverse as sports and pornography.

We argue that for both the US and Australian cases there is a concerted need to keep legacy journalism alive through traditional print organizations. This does not mean that these organizations should not change to embrace the new modes of conversation and participatory content creation. In fact, these institutions will be stronger for doing so. Further, we argue that we need to reassess what it means to be a journalist—seeing a place for both professional journalists and amateur journalism.

If the *New York Times* and Fairfax Media can embrace the changing terms of who can be a journalist, they may be able to find new ways to build audience loyalty by bringing their readers into the process of news creation. Journalism scholar Geneva Overholser (2008) has noted that in the present Internet age, being a journalist takes on new meanings as "curator, aggregator, news-recommender, beat blogger, community host, finder, network reporter, information architect, database manager, programmer, developer, group filter, sensemaker" (p. 105). Legacy news media may need to come to terms with the idea that there is no longer one kind of journalism—and they are no longer the only people practicing journalism—if they wish to be part of the broader conversation about news in the public sphere. The challenge for these legacy organizations is to find ways to argue for their own relevance as original content generators and find new ways to support this journalism. We may no longer see newspapers in their current form—as print papers with online components—but as we have shown, there is a significant need in both the United

States and Australia for the kind of journalism that newspapers provide for the functioning of democracy. Thus, the question we continue to ask as journalism changes and adapts to the online environment, is, who will provide the sort of journalism which will strengthen our democracy in the digital era? If we can return vigilantly to this question, then we can be hopeful that journalism will be able to meet both the challenges and opportunities it faces.

REFERENCES

A graphic history of newspaper circulation over the past two decades. (2009, October 26). *The Awl.* Retrieved from http://www.theawl.com/2009/10/a-graphic-history-of-newspaper-circulation-over-the-last-two-decades

ACMA Communications Report (2007-2008): Australian Communications and Media Authority. Retrieved from http://www.acma.gov.au/webwr/_assets/main/lib310777/complete07-08_comms_report.zip

Anderson, B. (1991). Cultural roots. In B. Anderson (Ed.), *Imagined communities: reflections on the origin and spread of nationalism* (2nd ed., pp. 9-36). London: Verso.

Annual reports, audited financial statements, and Form 990s. (2010). NPR.org. Retrieved from http://www.npr.org/about/privatesupport.html

Arango, T. (2009, April 28). Tentative nod to pay cut at the Times. *The New York Times.* Retrieved from http://www.nytimes.com/2009/04/29/business/media/29times.html

Australian Press Council (2008). *State of the news print media in Australia.* Retrieved from http://www.presscouncil.org.au/snpma/snpma_index.html

Beecher, E. (2009a, August 17). Death of newspapers: it's the advertising, stupid. *Crikey.* Retrieved from http://www.crikey.com.au/2009/08/17/death-of-newspapers-its-the-advertising-stupid/

Beecher, E. (2009b, September 30). Does Fairfax need a 67-year-old grocer at the helm? Crikey. Retrieved from http://www.crikey.com.au/2009/09/30/does-fairfax-need-a-67-year-old-grocer-at-the-helm/.

Benkler, Y. (2006). *The wealth of networks: How social production transforms markets and freedom.* New Haven: Yale University Press.

Brown, M. (2005). Abandoning the news. Carnegie Reporter 3, 1-5. Retrieved from http://carnegie.org/publications/carnegie-reporter/single/view/article/item/124/

Carey, J. (1969/1996). The communications revolution and the professional communicator. In E. S. Munson & C. A. Warren (Eds.), *James Carey: A critical reader* (pp. 128-143). Minneapolis: University of Minnesota Press.

Coming soon? The no newspaper city. (2008, December 13). Fitz and Jen. Retrieved from http://www.fitzandjen.com/2008/12/coming-soon-the.html

Curran, J. (2005). What democracy requires of the media. In G. Overholser & K. H. Jamieson (Eds.), *The press* (pp. 120-140). New York: Oxford University Press.

Davies, A. (Producer) (2008, August 28). The axe falls at Fairfax [Radio broadcast]. In Media Report. Melbourne: ABC Radio National.

Dikeos, T. (Reporter) (2009, May 26). Uncertain future for newspapers [Television series episode]. In B. Hawke (Executive Producer) The 7:30 Report. Australian Broadcasting Corporation.

Downie, L., & Kaiser, R. G. (2003). *The news about the news: American journalism in peril.* New York: Vintage.

Downie, L., & Schudson, M. (2009, October 19). *The reconstruction of American journalism. Columbia Journalism Review.* Retrieved from http://www.cjr.org/reconstruction/the_reconstruction_of_american.php?page=al

Fairfax Media (2009). *Corporate profile.* Retrieved from http://www.fxj.com.au/corporate-profile/corporate-profile.dot

Flew, T. (2008). Not yet the internet election: online media, political commentary and the 2007 Australian federal election. *Media International Australia incorporating Culture & Policy*, 126, 5-13.

Gillmor, D. (2004). *We the media: Grassroots journalism by the people, for the people.* Retrieved from http://oreilly.com/catalog/wemedia/book/index.csp

Glaser, M. (2006). Your guide to citizen journalism. *MediaShift.* Retrieved from http://www.pbs.org/mediashift/2006/09/your-guide-to-citizen-journalism270.html

Hirschorn, M. (2009, January/February). End times: Can America's paper of record survive the death of print? Can journalism? *The Atlantic.* http://www.theatlantic.com/doc/200901/new-york-times

HitWise. (2010). Top Australian Websites. Retrieved from http://www.hitwise.com/au/datacentre/main/dashboard-1706.html

Jarvis, J. (2009, May 12). Getting past newspaper's past. BuzzMachine. Retrieved from http://www.buzzmachine.com/2009/05/12/getting-past-the-past/

Kelly, J. (2008). *Pride of place: Mainstream media and the networked public sphere.* Cambridge, MA: Berkman Center for Internet & Society at Harvard University.

Kovach, B., & Rosenstiel, T. (2007). *The elements of journalism: What newspeople should know and the public should expect.* New York: Random House.

Lasica, J. D. (2007). What is participatory journalism?. *Online Journalism Review.* Retrieved from http://www.ojr.org/ojr/workplace/1060217106.php

Este, J., Warren, C., Connor, L., Brown, M., Pollard, R. & O'Connor, T. (2008). *Life in the clickstream: the future of journalism.* Retrieved from http://www.alliance.org.au/documents/foj_report_final.pdf

Lyons, D. (2009, September 27). Don't bail out newspapers -- Let them die and get out of the way. *Newsweek.* Retrieved from http://blog.newsweek.com/blogs/techtonicshifts/archive/2009/09/27/don-t-bail-out-newspapers-let-them-die-and-get-out-of-the-way.aspx

McCombs, M.E. (2004). *Setting the agenda: The mass media and public opinion.* Malden, MA: Polity.

Mnookin, S. (2008, December). *The New York Times'* lonely war. *Vanity Fair.* Retrieved from http://www.vanityfair.com/politics/features/2008/12/nytimes200812

McChesney, R.W. & Nichols, J. (2009, October 30) Yes, journalists deserve subsidies too. *The Washington Post.* Retrieved from http://www.washingtonpost.com/wp-dyn/content/article/2009/10/22/AR2009102203960.html

Nichols, J. (2007). Still a powerful voice. In C. M. Madigan (Ed.), -30-: *The collapse of the great american newspaper* (pp. 175-189). New York: Ivan R. Dee.

NPR Reaches New Audience High. (2009, March 24). NPR.org. Retrieved from http://www.npr.org/about/press/2009/032409.AudienceRecord.html

Overholser, G. (2008). Updating "on behalf of journalism: A manifesto for change." In *New Models for News: The Breaux Symposium* (96-111). Baton Rouge: The Manship School of Mass Communication.

Perez-Pena, R. (2009a, September 15). Times names Jonathan Landman culture editor. The New York Times. Retrieved from http://www.nytimes.com/2009/09/16/business/media/16times.html

Perez-Pena, R. (2009b, March 9). Times Co. building deal raises cash. *The New York Times.* http://www.nytimes.com/2009/03/10/business/media/10paper.html

Perez-Pena, R. (2009c, October 19). Times says it will cut 100 newsroom jobs. *The New York Times.* http://mediadecoder.blogs.nytimes.com/2009/10/19/times-says-it-will-cut-100-newsroom-jobs/

Pickard, V., Stearns, J., & Aaron, C. (2009). *Saving the news: Toward a national journalism strategy.* Washington, DC.: Free Press.

Rainie, L. (2009, November 13). The new news audience. *Pew Internet and American Life Project.* Retrieved from http://www.pewinternet.org/Presentations/2009/50--The-new-news-audience.aspx

Ricketson, M. (2008, August 27). Profit one week, job cuts the next. *The Age.* Retrieved from http://business.theage.com.au/business/profit-one-week-job-cuts-the-next-20080826-4330.html

Rosen, J. (2006, June 27). The people formerly known as the audience. *PressThink.* Retrieved from http://journalism.nyu.edu/pubzone/weblogs/pressthink/2006/06/27/ppl_frmr.html#more

Salmon, F. (2009, September 11). One problem with newspaper micropayments. *Reuters.* http://blogs.reuters.com/felix-salmon/2009/09/11/one-problem-with-newspaper-micropayments/

Schudson, M. (2003). *The sociology of the news.* New York: Norton.

Schudson, M. (2008). *Why democracies need an unlovable press.* Malden, MA Polity

Shirky, C. (2009, March 13). Newspapers and thinking the unthinkable. *Shirky.Com.* Retrieved from http://www.shirky.com/weblog/2009/03/newspapers-and-thinking-the-unthinkable/

Sparrow, J. (2009, August 14). Why we need a public newspaper. *New Matilda.* Retrieved from http://newmatilda.com/2009/08/14/why-we-need-public-newspaper

Stark, C. (2009, October 27). Disastrous newspaper circulation numbers, San Francisco Chronicle loses 25 percent of its readers. Star Silver Creek. Retrieved from http://www.starksilvercreek.com/2009/10/disastrous-newspaper-circulation-numbers-san-francisco-chronicle-loses-25-of-its-subscribers.html

Starr, P. (2009, March 4). Goodbye to the age of newspapers (hello to a new era of corruption). *The New Republic.* http://www.tnr.com/politics/story.html?id=a4e2aafc-cc92-4e79-90d1-db3946a6d119&p=2

State of the News Media (2007). Pew Project for Excellence in Journalism.Washington, DC.

State of the News Media (2009). Pew Project for Excellence in Journalism. Washington, DC.

Times ponders online fee$. (2009, 12 September). *NY Post.* Retrieved from http://www.nypost.com/p/news/business/times_ponders_online_fee_VvZriCGVJIInoTl2oA9NPJ

Top 15 most popular news Websites - January 2010. (2010, January). *Ebiz MBA.* Retrieved from http://www.ebizmba.com/articles/news-websites

Westphal, D. (2009). *Philanthropic foundations: growing funders of the news.* Los Angeles: University of Southern California: Center on Communication Leadership & Policy.

Usher, N. (2009, December 22). The business model for news is and always has been broken and Rupert Murdoch can't fix it. *Online Journalism Review.* Retrieved from http://www.ojr.org/ojr/people/nikkiusher/200912/1808/

Zappone, C. (2008, August 26). Fairfax Media to cut 550 jobs. *The Age.* Retrieved from http://business.theage.com.au/business/fairfax-media-to-cut-550-jobs-20080826-42fu.html?page=fullpage#contentSwap1

Zelizer, B. (2005). Definitions of journalism. In G. Overholser & K. H. Jamison (Eds.), *The Press* (pp. 66-80). New York: Oxford University Press.

SECTION TWO

TECHNOSOCIAL INTERSECTIONS

CHAPTER FIVE

Media Literacy in the Facebook Age

Designing Online and Face to Face Learning Environments

ANDRÉS MONROY-HERNÁNDEZ, MICHAEL DEZUANNI, &
KAI KUIKKANIEMI

This chapter explores how young people develop media literacies in learning environments that have different levels of formality and informality. Media literacy is an educational approach that aims to enhance young people's knowledge about media and media production skills to enable them to productively participate in a range of social and cultural contexts. The theorization of media literacy in this chapter moves beyond some established approaches that aim to develop young people's critical reading skills. Instead, the chapter recognizes that young people are often media content producers and that media literacy is developed socially and culturally in an ongoing fashion. Formal schooling is just one setting in which media literacy can be developed and it should be seen as much more than the attainment of an educational goal or competency. Indeed, young people participate in many communities in which they use and develop media literacies. This chapter discusses three environments that allow young people to produce media using new media technologies, in particular video games and digital animation. In recent years, the benefits of using new media forms for educational purposes have been well established. One environment discussed in this chapter, the Video Games Immersion Unit, was constructed in a school setting and includes a blend of online and face-to-face experiences, while the other two environments, the Scratch online community and the Habbo online world are entirely nonschool online experiences. These examples (which are introduced in more detail in section "The Learning Environment") are discussed to identify the opportunities they provide for students to develop media literacy skills and knowledge. We have identified four common characteristics that are important for the development of media literacies across the three distinct environments: peer learning ("Peer Learning in the Three Environments" section), mentoring ("Mentoring in the Three Environments" section), using technological tools in unexpected ways ("Unexpected or Novel Uses of the Tools in the Three Environments" section), and establishing reputation ("Establishing Reputation in the Three Environments" section). Each of these is discussed separately and then the connections between them are identified.

The media literacy field is undergoing significant changes due to the evolving nature of the relationship between young people and media. In the past, media literacy education focused on providing young people with the skills to decode or analyse media texts (Leavis & Thompson, 1933; Masterman, 1980, 1985; Thompson, 1973). This was based on the assumption that young people required critical analytical skills to meaningfully participate in media cultures and to avoid being unduly influenced or exploited by powerful media. More recent theorizations of media literacy, for example, by Jenkins (2006), Ito (2010), Ito et al. (2008), Livingstone, Van Couvering, and Thumim (2008), and Buckingham and Domaille (2009) emphasize the development of young people's critically reflective social participation in media cultures and recognize that young people are not deficient in their relationships with media, but are active and proficient participants. This has gained impetus with the availability of new media technologies that allow young people to easily produce their own content and socialize in online spaces. Jenkins theorises this as "participatory culture":

> Participatory culture shifts the focus of literacy from one of individual expression to community involvement. The new literacies almost all involve social skills developed through collaboration and networking. These skills build on the foundation of traditional literacy, research skills, technical skills, and critical analysis skills taught in the classroom. (Jenkins, 2006, p. 4)

The social media literacy skills required for productive social and cultural participation in media cultures have been theorised through complementary but different perspectives. For example, Buckingham (2007) argues that young people require a conceptual framework for reflecting on their production and use of media that includes asking questions about the languages used to communicate with media, the representations of people, places, and ideas constructed through media, the audiences for whom media are made, and the institutional contexts within which they are produced. Ito (2010) asks what skills and knowledge are necessary for young people to move from "hanging out" and "messing around" with media (e.g., through participation in social network sites) to "geeking out" with media (being more productive with creative technologies). Jenkins (2006) outlines 11 skills that are required for the development of new literacies for successful participation in media cultures: play, performance, simulation, appropriation, multitasking, distributed cognition, collective intelligence, judgement, transmedia navigation, networking, and negotiation. In this chapter, we take a different approach and identify the *design features*, or *affordances*, necessary in environments to provide opportunities for young people to develop media literacies at a range of knowledge and skill levels. We argue that unless these features are present, it is unlikely that young people will develop the skills and knowledge identified by Buckingham, Ito, and Jenkins.

We discuss three different social environments where young people engage in developing new media literacy skills through the lens of four phenomena: peer learning, mentoring, unexpected uses of digital tools, and the development of reputation. The Video Games Immersion Unit was designed collaboratively by media and technology educators as a specific educational experience in a school environment. The Scratch online community was designed as an educational space in which students can share their creative production work. Habbo was designed as a space for socializing, but has educational implication. Our analysis of these three environments is a grounded approach in which we aim to identify specific examples of students' use of the environments for creative and social interaction. All three environments involve young people in play and work with video games and digital animation, which are significant media forms in young people's lives. Each of the spaces has been designed to encourage young people to interact socially and to be creative. In the case of the Video Games Immersion Unit and the Scratch online community, education was the underlying design objective. Habbo Hotel was designed for socialization and entertainment. On a continuum, the Video Games Immersion Unit is the most formal learning environment, while Habbo Hotel was the least formal.

Each of the three environments creates a form of community that enables forms of peer learning. Peer learning includes collaboration, teamwork, and shared problem solving. Mentoring was also evident in each of the environments. When young people mentor one another, they provide each other with alternative ways of solving problems and understanding processes from different perspectives. Mentoring is therefore a form of peer teaching and is a common feature of distributed networks, where less emphasis is placed on a few individuals holding knowledge and distributing it to many (as is the case in traditional school classrooms). For example, Ito et al. (2008) identify how children sometimes mentor their parents in new media environments. The establishment of "reputation" is important within all three environments and involves young people developing expertise for which they are recognized. Developing reputation is an aspect of the ongoing development of identities which are crucial to learning in new media environments. For example, Gee (2003) argues that young people learn in video games environments through taking on "projected identities." Each of the environments also allows for the unexpected use of digital tools. This is a crucial affordance that allows for experimentation, play, and creativity, which are central to the development of media literacy in new media environments. According to Jenkins (2006, p. 4), "the ability to meaningfully sample and remix media content" is a key skill for participation in new media cultures. Each of these affordances is discussed in greater detail in the analysis sections that follow.

THE LEARNING ENVIRONMENTS

The Scratch Online Community

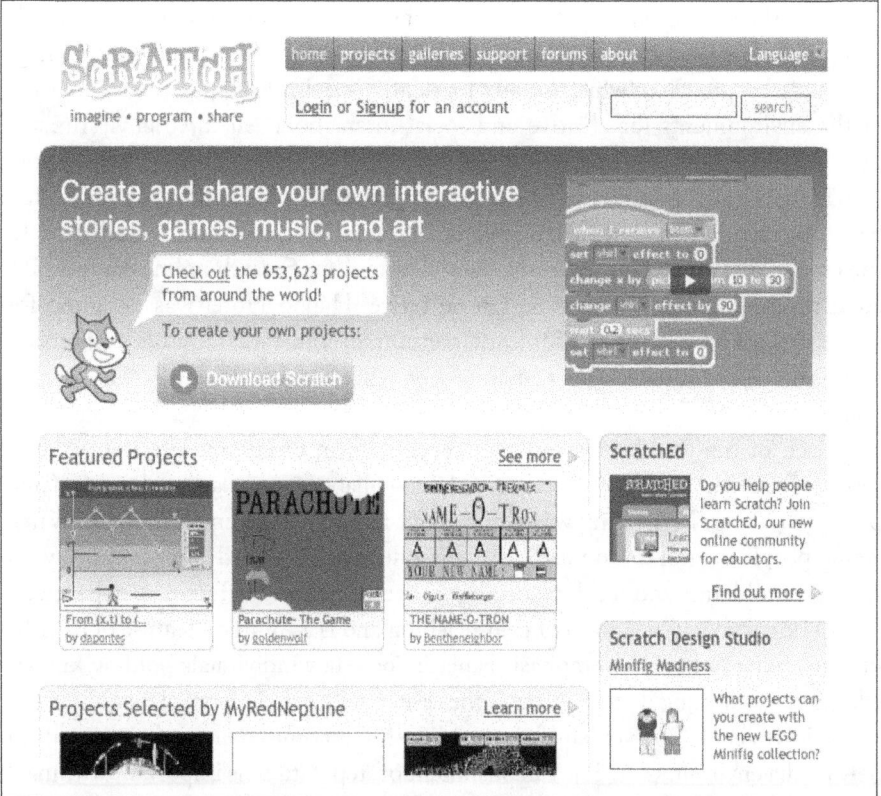

Figure 1. Home page of the Scratch website, January 2010

The Scratch website[1] (figure 1), dubbed the "YouTube of programming," is an online community where young people share their own video games, animated stories, and interactive art (Monroy-Hernández & Resnick, 2008). Members of the Scratch online community use the Scratch programming environment, developed by the Lifelong Kindergarten group at MIT, to program their digital artifacts by putting together blocks of code to control the interactions of visual objects and sounds (Resnick et al., 2009). Two and a half years after its release in 2007, more than 800,000 projects have been shared on the Scratch website. Projects range from physics simulations, to video games featuring Obama and McCain, to animated stories of singing cats.

Every month, more than half a million people[2] from around the world visit the Scratch website. There are more than 400,000 registered members, and 25% of them have shared a project. The vast majority of users are between 8 and 17 years old (self-reported), and there is an active minority of adults who often play the role of mentors.

People use the website not only to share their work but also to interact with other creators, exchange ideas, work on collaborative projects, and discuss their daily lives. A number of collaborative efforts have succeeded in creating dozens of projects in what kids often refer to as "companies"; that is, a group of kids who cocreate projects. The website is completely open: anyone can browse, download, and interact with people's projects, or register for an account to post their own. Participants are encouraged to download other people's projects to learn how they were created and reuse parts to create remixes; in fact, 28% of the projects are remixes. All projects are shared under a Creative Commons license. Registered members can tag, "love,"[3] and bookmark projects. Furthermore, in the spirit of popular online social networks such as Facebook, they can befriend other creators while maintaining the main goal of creating projects.

The Scratch project has helped in fostering new media literacy by providing kids with the tools and the social environment to become full participants in the creation of digital culture. This achievement is exemplified by a message from FlashTide,[4] a 12-year-old boy who sent a message to the creators of the website:

> Dear Scratch Team, I don't know if this is the case, but besides being a place to share, I think you planted this website as an experiment: you might have been wondering how a 4-kids social networking site would develop. Well, develop it did! If you're new to Scratch and you look around, you can't help but notice that there are 'celebrities' out there like Nicole12, TheWizard, and JohnDoodle. The featured, Loveit, and top viewed sections let lesser famed people get the spotlight too, and the creation of companies (I would know, I'm in one) such as DG games, Moo productions (yay) and countless others. If this website were, in fact, an experiment, I'd think it was a successful one! (Message received by the Scratch Team via the "Contact Us" web form, December 2009)

Video Games Immersion Unit

The Video Games Immersion Unit was a specialized program developed for a Catholic boys' school upper middle program in Brisbane, Australia. It involved a group of 14 to 16-year-old students who studied and produced video games in an intensive mode for 15 days. The Immersion Unit aimed to combine media literacy and technology education objectives. The media education specialist (Dezuanni) aimed to have students become more critically reflective about video games, while the technology education specialist aimed to have students learn specific software skills and technological processes (for further description of the Unit content see Dezuanni, 2010). Learning in the Immersion Unit occurred in a "blended" environment, as learning took place in both face-to-face and online modes. The face-to-face interactions occurred in a computer laboratory, which was converted to include several gaming stations, in addition to the computer terminals. Online learning occurred in a MOO space (see Figure 2) in which students constructed their own personal and team "rooms," which they could decorate in any way they wished. Students performed many tasks in this space and could collaborate and share ideas. The space included tools for blogging, wiki style collaboration, video and image sharing, forums, and synchronous chat.

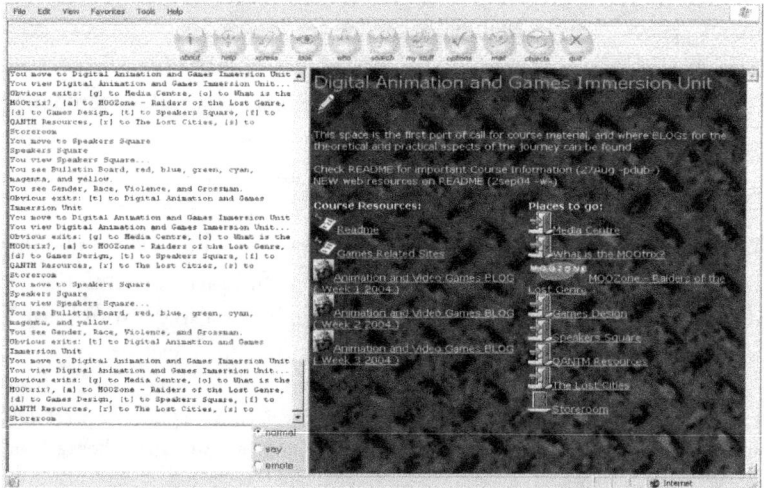

Figure 2. The Video Games Immersion Unit MOO space

Some aspects of the program were relatively formal. For example, a number of sessions involved didactic instruction, particularly for training in software skills, and students were required to complete a collaborative assignment for assessment purposes. There was a planned structure to the unit, with specific tasks to be completed in broad time frames. However, other aspects of the program were more informal. In the context of this private boys' school, freedoms such as a lack of specific lesson times, not having to wear a uniform and casual interaction with the teachers were experienced by the students as informalities. Furthermore, students were encouraged to play games at their leisure, between work sessions. There were significant periods of time when students were able to work at their own pace on production tasks. The online environment also allowed a great deal of after school collaboration during which students continued the days' work, generally without the presence of teachers.

Habbo Hotel

Habbo is a massive graphical social media website, with isometric 3D graphics (see Figure 3). It was launched in 2000 and has expanded to include 31 online communities or "hotels." In practice, every hotel is located in a different county. According to Sulake Corporation (the company that owns and operates Habbo) there are over 160 million registered avatars in Habbo, and 16.5 million unique visitors monthly. Ninety percent of Habbo users are between 13–18 years old, and according to Lehdonvirta, Wilska, and Johnson (2009), one quarter of Finnish teens (aged 13–18) have registered to use Habbo. Each Hotel has unique characteristics, paid "hotel managers" for content moderation and special localized events such as celebrity visits, competitions and marketing campaigns. Sulake Corporation has paid special attention to social responsibility within Habbo, and tried to make it a safe place for teens to use. The corporate and Habbo website have a visible content category called "safety," and in addition to automatic chat moderation and spending limits, there are paid employees managing discussion and eliminating misbehavior. Each user has a Habbo avatar and Habbo room, and can easily change the appearance of their avatar. Decorating a room requires items, which can be acquired by user-to-user trading or purchasing from Sulake. Habbo also has nongraphical components, including a website which contains user pages, group sites, event lists, general information, and forums. In addition, there is an official Habbo fan site and a Habbo toolbar.

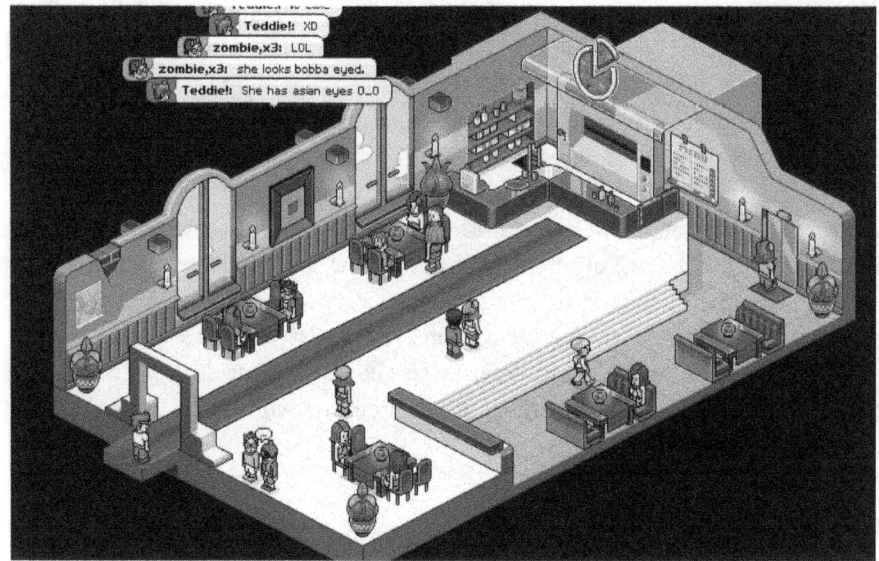

Figure 3. A public room in Habbo, February 2010

PEER LEARNING IN THE THREE ENVIRONMENTS

The Scratch Online Community

Members of the Scratch online community engage in peer learning in implicit and explicit ways. Implicitly, by downloading other people's projects, people learn project type specific programming techniques. For example, the user known as "TheWizard" is well known in the community for having popularized a type of game that uses the "scrolling background" technique. Knowing this, TheWizard created a template project that clearly demonstrates the technique. Members of the Scratch community also use the discussion forums on the websites to ask for help or offer their expertise, constituting an explicit peer learning practice. For example, a user created a thread title "Ask me to help you with Program!" where he offers his help to new users: "I will help you, new user! I know a lot thing, like, Velocity or Scrolling even something Great!!!! It could be 3-D or it could be 2-D! :-D !!! Comment, to let me know if you need any help! Add your problem on last of comment! :-)" (Post on the Scratch forums, December 2009, *sic*)[5]

Another user offers a similar service and even goes as far as to give the promise that if she does not know the answer to the question she will find someone who does: "I got your back. I can help you, and if you aren't satisfied? I'll get you someone else! Just post below!" (Post on the Scratch Forums, June 2009.) Other users ask for help, for example, this user explains a technical problem she is dealing with a program she is having:

> Currently I'm working on a new project and so far is going very well..... But there's one problem I'm trying to make a sprite glide across the screen—but if it comes in contact with another sprite in doing so, the sprite will stop (and if possible wait for the sprite to pass before continuing on gliding across the screen) Any help will be greatly appreciated!!!!! (and yes i am a noob when it comes to anything Scratch...)" (Posted on the Scratch Forums, October 2009).

A few hours later a couple of people responded, one of the responses (Figure 4) was what she was looking for, to which she replied:

> "Thank you!!! thank you !!! thank you !!!
> I JUST CAN THANK YOU ENOUGH!
> I tried what you said and it worked !!! xxxxxxDDDDDD" (ibid.).

Interactions like this one happen every day on the Scratch website, however, knowing how to have the social and technical infrastructure to support them as the community grows continues to be a challenge.

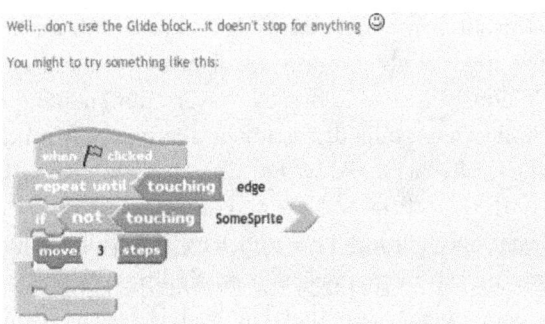

Figure 4. Detailed response to a question on the Scratch Forums

Video Games Immersion Unit

Peer learning was a central aspect of the Video Games Immersion Unit. The unit designers deliberately aimed to create a sense of community amongst the class as a whole and more specifically through the creation of learning teams. The unit structure required students to work in teams to produce a concept for a video game and then produce aspects of the game using the software they received training in. Student blog reflections written after each learning episode revealed that they were motivated to work with their peers and that they gained a great deal of pleasure from this. For example, at the end of the first day, a student reflected: "The first day was great fun we got into groups of four, Link, Woodstock, rbimdxe and me. The experience that I will undertake over the next three weeks will be an exciting and fun one because of the people who are doing it and my interest in the topic itself" ([The One] blog reflection).

It was clear that the students also gained a great deal of pleasure and reward from the implication that their teams approximated professional gaming companies. For example, another student reflected: "Today was awesome. It felt like as if we were a real gaming company by the fact that we had to develop a game before a deadline, discuss our work with our senior heads (our parents), and having to attract a target audience into our game design and concept" ([Woodstock] blog reflection). [Woodstock] reconstructed collaborative problem solving with his peers as an opportunity to perform aspects of video games professional practice. In this context, peer learning is an enthusiastic rehearsal for imagined "real world" work and there is a playful engagement with the processes of game production that [Woodstock] constructs as social and competitive.

Habbo Hotel

Habbo has many features that allow users to communicate and share ideas. Many of the private rooms are constantly accessible and users can learn how others create rooms and use items. Public rooms are always open, and graphical chatting metaphors enable non-intrusive lurking to conversation, making it easier for new users to observe and join ongoing discussion and activities. Mimicking is an effective way of learning the behavior and system affordances in Habbo. Outside Habbo's graphical space, both the official website and fan sites are important resources for all kinds of information regarding how to behave in and use Habbo. Overall, the multitude of communication channels and possibilities for participation, most of which are easily accessible, means that Habbo is a relatively open and creative environment for teens.

MENTORING IN THE THREE ENVIRONMENTS

The Scratch Online Community

Mentors have emerged organically in the Scratch community. Through their projects and interactions during a long period of time, some members gain a lot of respect from the rest of the community. Due to their consistent helpful behavior, some of these users have been selected by the administrators of the site to become forum moderators. Among them is "Paddle2See," a retired engineer who was introduced to Scratch by his son: "I have tinkered with 'Recreational Programming' for years, [...] so when my youngest son came across Scratch on a computer at his High

School, he knew I would be interested and he showed it to me" (Paddle2See, via e-mail interview, February 2010). After months of participating in the community people noticed Paddle2See's expertise and asked him for advice:

> As I got more involved with the community, I would come across a lot of new (and not so new) Scratchers that seemed to be working in isolation. They would post up their creations and get little or no feedback. I found that disturbing...everybody needs encouragement and feedback (especially children!). So I got into the habit of spending a lot of my spare time just cruising the New Projects channel and commenting on projects. That lead to trying to help them with programming issues and, occasionally, social issues as well. (ibid.)

About a year later, Paddle2See was a well-known member of the community and was invited to attend the first Scratch conference at MIT. After meeting him face-to-face, the administrators of the website decided to ask him to join the staff of moderators. Paddle2See continues to be a role model for a lot of kids, offering helpful advice, contributing to the moderation of the forums, and helping with the selection of projects for the front page. This story exemplifies the mentoring role educators and thoughtful adults can play in online communities.

Video Games Immersion Unit

There was a constant process of peer mentoring throughout the Video Games Immersion Unit, particularly while the students were learning software skills. This occurred spontaneously in the sense that the teachers did not need to ask the students to help each other to solve problems. Students who quickly acquired skills often took it upon themselves to assist other students to complete tasks, as the following student explains:

> Today we started to learn action scripting. After we had spent about half an hour listening carefully to the instructions from our instructor we set to the task of giving our characters keys to use their attacks. Luckily I received no errors the entire day in action scripting and had to help a few others with figuring theirs out. ([Rbimdxe] blog reflection)

This occurred particularly within teams, but also between members of different teams of students. Indeed, peer mentoring was such an expectation throughout the Immersion Unit that when students were denied the opportunity to help each other, they showed a great deal of less satisfaction with the learning process. For example, for 3 days of the Unit, the students undertook intensive Flash animation training at a multimedia training facility, which involved a form of lock-step training in which an instructor demonstrated a skill the students were then required to emulate. The pace of this training meant that students mostly had to concentrate on completing their own work, with very little opportunity to help their friends.

Several students complained about the difficulty of the work, particularly that they were falling behind in their understanding. One student reflected that he "was simply a sheep...clicking on buttons, blindly following the set instructions for my simple task in life" ([Rogue Splitter] blog reflection). The lack of social learning while learning Flash resulted in the highest number of negative student reflections about any activity completed throughout the 3 weeks of the Immersion Unit.

Habbo Hotel

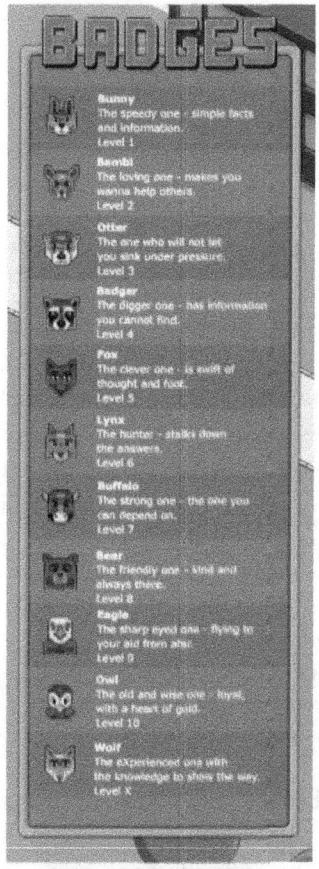

Figure 5. Habbo Badges

Since the creation of Habbo, there have been several different official systems for mentoring. The first system (from 2001–2005) was called Hobbas. This is how one experienced user described Hobbas: "Hobbas were very much like mini-moderators for the hotel, they were volunteery members to make sure people were not breaking the rules, they also educated Habbos with how to be safe, and guided them around the hotel." (Post on user Agesilaus).

Hobbas had special badges and special rights in the system. However, due to the expansion of the hotels and misbehavior on the part of some Hobba's, Sulake decided to cancel the Hobba program and replace it with HabboX mentors and paid employees. HabboX were experienced users with fewer special privileges than Hobbas. Ultimately the HabboX was also cancelled and changed to Habbo Guides. Figure 5 shows what kind of roles Habbo Guides can have and how they can advance.

Johnson (2007a) has developed Habbo user profiles based on cluster analysis. There are six user categories: oldtimers, playmakers, silent majority, gang-members, I don't pay players, and older people. Playmakers, who make up to 15% of the total, are defined as: "we like to visit often and arranged events for others." This is a group of people who can be considered mentors in Habbo and who seem to play a very important role in supporting the growth and culture of the community, something that it is often ignored to happen in informal environments like Habbo.

UNEXPECTED OR NOVEL USES OF THE TOOLS IN THE THREE ENVIRONMENTS

The Scratch Online Community

A sign of success in any tool and social environment is their use in unexpected ways. As such, it is particularly encouraging to see both Scratch's tools and social spaces being used in novel ways. One example of this is 10-year-old 'yodaboys' science fair project, the "Reflex Tester" (Figure 6). The idea was to use the Scratch website as a crowdsourcing tool to get participants to perform a reaction time experiment programmed in Scratch. The project displayed a button that people were asked to push whenever the color of the background changed. At the end, the program would show the reaction times. Yodaboy asked participants to post these times as a comment on the project, along with an answer to the questions related to sports playing (such as "Do you play a sport," "Which one?" and "How many hours a week?"). These results were used to determine if there was a connection between playing sports and reaction times. More than 300 people answered the survey, with the results posted by yodaboy as a separate project. This project won the science fair competition at yodaboy's school, and he probably learned much about the power of social computing even if he might not use the term itself. This example represented a breakthrough in the use of Scratch and the website—they were no longer two separate entities, but became one single tool through yodaboy's specific use.

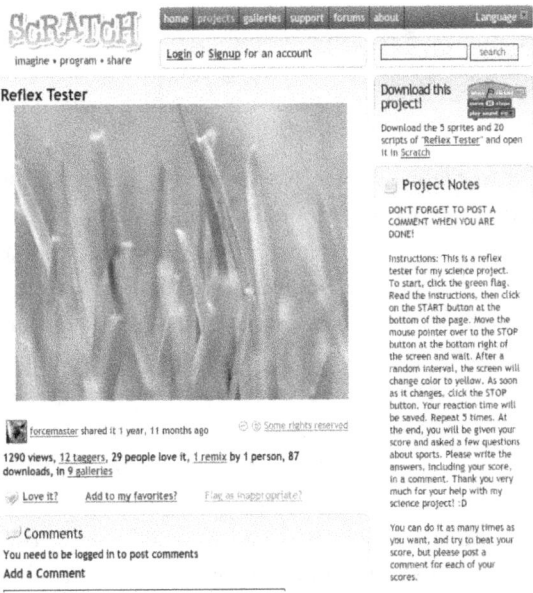

Figure 6. Project of user exploring crowdsourcing

Video Games Immersion Unit

The formal aspects of the Video Games Immersion Unit somewhat restricted the extent to which students were able to use the software tools in unexpected or novel ways. For example, they were required to produce game artifacts for the concepts they developed in their teams. However, the ways in which they achieved this through forms of experimentation and play with digital tools demonstrated high levels of technological competence and digital literacy. For example, the following is a brief excerpt from a long explanation of the technological process of problem solving to make a games artifact from one student:

> I first made the blade, by making a square-based pyramid, stretching it at the tip, then squished it. After that I made a hilt by combining two square-based pyramids (one that is upside down) and chopping some bits off by using positives and negatives. Trying to make complicated objects by using Bryce 5 was a challenge, so I tried using wings3D, but it took too long to load, so I used Bryce 5 after all. ([Link] blog reflection)

Experimentation and play with digital technologies was a constant feature of the students' experience of the Video Games Immersion Unit, and this was a deliberate design feature built in by the Immersion Unit teachers. In the majority of cases, students were shown how to use a software technology and were then expected to use it in their own way to complete the task of producing aspects of their video games. Even when the students learnt Flash animation through a very didactic process, they were eventually able to experiment with the technology and gain a great deal of pleasure and satisfaction from the process. It was obvious to the teachers that experimentation with the technologies was crucial to the learning that occurred throughout the Immersion Unit.

Habbo Hotel

The development of Habbo has been significantly impacted by user innovation. Based on user and developer interviews, Johnson (2007b) has listed several formal Habbo features that were originally inspired by users. Examples of such inspiration include trading units, Habbo professions, games based on simple Habbo items, and user cooperatives. The purchase of virtual items is the core business model for Habbo. Initially the system did not support comprehensive trading between users. Based on user innovation (Lehdonvirta, 2009), users developed an unofficial trading system based on plastic chairs. Plastic chairs and other value references were used to define the price of other more valuable items. Later on, Habbo added several features to ease the user trading, such as secure trading client and exchange coins. Habbo users have been very innovative in creating activities by themselves. Figure 7 shows a room, which is basically a bingo game, based on very simple dice furniture component.

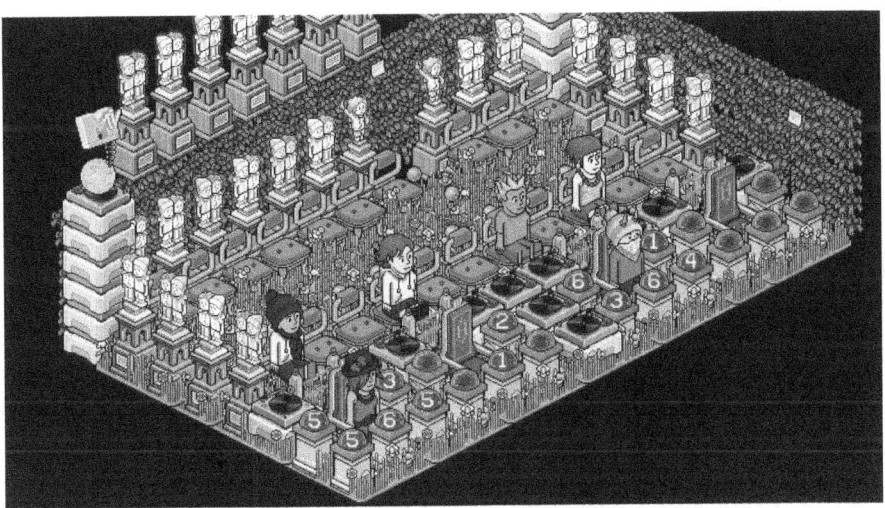

Figure 7. Bingo room in Habbo, screenshot from habbo.fi, February 2010

Sulake is very proud of user-generated content (UGC) in Habbo. According to their website there are several types of UGC in Habbo:

> Check into Habbo Hotel and explore the different user-generated theme rooms, such as party rooms, maze rooms, barbershops, police stations, hospitals, cruise ships, gang centers, market places, adoption centers and whatever you can imagine—and often what you haven't even thought of (Sulake, 2010).

This text is prominently placed on their corporate website and demonstrates both the importance of UCG for Habbo and the fact that user innovation in Habbo often goes beyond even Sulake's own imagination.

One famous example regarding user innovation was a user collective called Cruise Ship Cosmos, which was active in the Finnish Habbo from 2004 to 2006. According to Johnson (2007b), Cosmos linked together more than 100 user-created rooms. Each room matched exactly the style guidelines and obeyed commonly agreed behavior protocols. The collective had more than 350 partners and over 1000 regularly attending VIP customers, and many more infrequent customers. Participation in Cruise Ship Cosmos, which was organized by pre-teens and teenagers, required a financial contribution (in terms of purchased virtual items). The collective operated for several months without Sulake being aware of its existence. This is, to us, a striking example of how young people, given the freedom and the right structures, can develop complex social systems empowered by the right technological affordances.

These and the other examples show to us that successful technologies and systems that are easily repurposable and appropriated by young people are not only more authentic and successful at engaging people, but are also fertile ground for extending the uses of the tools to new ways of thinking and broader audiences with a diverse range of interests.

Establishing Reputation in the Three Environments

The Scratch Online Community

Building and maintaining reputation is one of the main driving forces for participation and conflict in the Scratch online community. The front page features different projects based on a combination of user-driven metrics and centrally managed ones. For example, the three most "loved" projects are presented on the front page along with a list of "3 featured" projects picked by the administrators of the website. Having one's project on the front page is considered one of the highest social rewards in the community. For example, a 14 year old whose project was displayed on the front page for a few days commented: "OH MY GOD! THIS IS ON THE TOP VIEWED LIST!!!!! [...] THIS IS THE HAPPIEST DAY OF MY LIFE!!!!!." (Comment posted by the creator on the description of her project after it got on the front page). This reputation incentive is used by the site administrators to promote certain types of behaviors. For example, the "Top Remixed" section was created to change the perceived value associated with remixing by increasing the reputation of users whose projects were remixed. The goal was to turn something perceived as negative, like "copying," into something potentially positive. Other participants have often used the numbers of "love its" (indications of support for a project) to gauge the attention their projects get. For example, a project creator posted the following on one of the notes next to a project: "if I get 5+ love its I will extend this game!" (Comment posted by the creator on the description of her project). Seeking reputation can also lead to a lot of conflicts. One of the most controversial examples was the case of a participant who created projects claiming that unless the project reached a certain number of "love its," a seemingly imaginary disease would cause the creator some kind of harm.

Video Games Immersion Unit

Establishing reputation played a significant role in the creation of social acceptance and effective learning in the Video Games Immersion Unit. Students strived to become recognizable to other students as good game players and as being proficient with technologies. Students who acquired skills and knowledge efficiently were admired by the other students and were often turned to for assistance and advice.

In the best cases, this occurred in a manner that led to distributed skills and knowledge so that some students became experts in a particular area, and others became experts in different but complementary areas. Of course, this did not always occur in one team, but across the whole class there was a productive level of diversification of skills and knowledge. Some students became highly skilled programmers, while others were more adept at design work or at planning and script writing. It was also clear that the students developed identities through the acquisition of skills (Dezuanni, 2010).

In one activity that relates to reputation from a critical media literacy perspective, and which had a focus on gender and video games, the students were required to take part in an online chat with university aged female gamers. Several of the students expressed initial concern about this activity and some male students rejected the female students' ideas on the basis that the female students did not seem to have adequate knowledge about games. However, there were also several instances of male students accepting the female students/gamers as authoritative. For example one student says:

> I almost always agreed with what Natasha had to say, as she had quite a knowledge of what was the purpose of video games and their relation with sexism. One of the things I learnt today was that just because Tomb Raider gave precedence to a main female character, it still didn't necessarily appeal to females. ([Rhysay] blog reflection)

In this case, establishing reputation became a point of contestation that presented an opportunity for students to broaden their knowledge about gender and games. Although not all students responded in a way that the teachers might have preferred, the opportunity to consider reputation in a critical way was important.

Habbo Hotel

According to Sulka Haro, the lead designer of Habbo, leaving Habbo for 2 months means that you need to spend a lot of time to catch up and build your profile back up again (Gamasutra, 2009). This means that reputation within Habbo is important but dynamic or nonpersistent. There are several ways to maintain reputation. There are official badges (like the guide badges shown in Figure 2), which are visible declarations of achievements. Based on user feedback, Habbo designed a respect system in 2008. The respect system allows users to directly give respect to other users. Furthermore, the popularity of user's rooms is an indicator of user reputation. There are also informal ways to gain reputation in Habbo. According to Johnson and Toiskallio (2005), many of the fan sites rank rooms based on reputation. Also, owning certain rare items can make you respectable within certain parts of the community. Item trading and item valuation is an important part of fan site content. Furthermore, Habbo is full of specialized and private communities, which can have very elaborate internal hierarchy and reputation systems. For example, in the case of Cruise Ship Cosmos, the user community defined over 42 roles, which were divided into seven main organizational levels. These details about the reputation mechanisms in Habbo help us surface that, even in very informal and mostly hedonic environments, a structured reputation-building mechanism are part of the user-experience that, as we see later, help cement people's interactions.

Discussion and Concluding Remarks

In this comparative analysis, we have focused on identifying commonalities among three real-world environments where young people engage in the use and creation of digital media. The commonalities are presented as a framework for designing similar social spaces aimed at promoting new media literacy skills. Based on the previous observations, we propose the following four considerations for designing online spaces:

First, we argue that allowing peer learning to emerge is of utmost importance and, based on our observations, we identified that direct channels for communication are key in supporting it. For example, both Habbo and the Immersion Unit had clear mechanisms for people to talk to each other, and even though the Scratch community is primarily about sharing interactive objects, participants can post comments on each others' work. These channels of communication also help create a culture in which peer learning is supported.

Second, we see mentoring as an emergent property of a system that allows for peer learning and reputation building mechanisms and as such, we encourage facilitators and designers to avoid "getting in the way" of these emergent community leaders. Facilitators and system designers can often act as matchmakers between mentors and participants looking for someone to teach them the different aspects of the tools or environment. Mentors also often bring a balance to the informal and formal spectrum by distributing control and helping establish a balance between the formal and the informal.

Third, we advance the idea that the socio-technical openness in the three environments helped support the use of the technologies in unexpected ways. These unexpected uses of the tools and environments are presented here as a positive sign that people feel free to appropriate these to participate in personally meaningful ways, explore their own identities, and engage in play and experimentation. While we strongly believe openness is a key factor in the success of the environments, we also advocate in favor of having certain structures that frame the interactions in specific ways and avoid compromising the core utility of the system.

We think of this as "sandbox" openness. For example, the Scratch online community focuses primarily on the sharing of animations and video games in an open way without prescribed templates, but at the same time it deemphasizes and makes it hard to have other types of participations, such as using the system as a chat room (without completely preventing it). Our main design suggestion here is to be open and receptive to unexpected uses of the technologies, but to still provide the right socio-technical infrastructure to frame and guide the interactions.

Fourth and finally, reputation-building mechanisms are identified as another important design element to foster collaboration and a productive media literacy learning environment. In both informal and formal ways, reputation emerges out of peer-to-peer interactions when participants have a way of forming an identity with history attached to it. However, there are ways in which a system and/or a facilitator could manage reputation in a destructive way, either by monopolizing it (as in some traditional classrooms where only the teacher can be the expert, or in online interactions where only the administrators of the system are in charge of broadcasting information), or by over-empowering a few and letting these individuals take over the communication channels. For example, during the week in which the Immersion Unit had an expert teach about a specific topic without allowing for peer learning, the class was less effective than when the free flow of information and peer-to-peer communication was permitted. These reputation mechanisms often allow people to emerge as leaders which, given the right circumstances, can significantly empower a community to participate and make it their own.

Inspired by the work of Jenkins, Ito, Livingstone, and others who have made arguably the best arguments to stress the importance of new media literacy skills, our initial analysis of three distinct environments has led us to propose a framework for thinking about the design of social and technical systems to support the development of these literacy skills. This framework lies on the idea that openness will lead to new and unexpected uses of technological affordances, and that this openness should also extend to the communication channels where peer learning can be fostered. Finally, we argue that adding the right mechanisms for building reputation will lead to the emergence of mentors. We hope that designers, facilitators, and educators find this framework productive when designing their own new media literacy learning communities.

Notes

1. The website is accessible at http://scratch.mit.edu
2. Web analytics available at quantcast.com/scratch.mit.edu (last accessed February 26, 2010)
3. There is a "love it" button on every page where projects are displayed that allows users to praise other people's projects.

4. There is a "love it" button on every page where projects are displayed that allows users to praise other people's projects.
5. All children's quotes, unless specified otherwise, are reproduced verbatim—any errors are source rather than transcription based.

REFERENCES

Buckingham, D. (2007). *Beyond technology: Children's learning in the age of digital culture*. Cambridge, UK: Polity.

Buckingham, D., & Domaille, K. (2009). Making media education happen: A global view. In C.-K. Cheung (Ed.), *Media education in Asia*. New York: Springer.

Dezuanni, M. (2010). Digital media literacy : Connecting young people's identities, creative production and learning about video games. In D. E. Alverman (Ed.), *Adolescents' online literacies : Connecting classrooms, media, and paradigms*. New York: Peter Lang.

Gamasutra (2009). GDC: Habbo's Haro talks marrying social worlds with game mechanics. Retrieved on February 28, 2010, from http://www.gamasutra.com/view/news/22861/GDC_Habbos_Haro_Talks_Marrying_Social_Worlds_With_Game_Mechanics.php

Gee, J. P. (2003). *What video games have to teach us about learning and literacy*. New York: Palgrave Macmillan.

Ito, M. (2010). *Hanging out, messing around, and geeking out : Kids living and learning with new media*. Cambridge, MA: The MIT Press.

Ito, M., Horst, H., Bittanti, M., boyd, d., Herr-Stephenson, B., Lange, P. G., Robinson, L. (2008). *Living and learning with new media: Summary of findings from the digital youth project*. Chicago, IL: The MacArthur Foundation.

Jenkins, H. (2006). *Confronting the challenges of participatory culture: Media education for the 21st century*. Chicago, IL: The MacArthur Foundation.

Johnson, M., & Toiskallio, K. (2005). Fansites as sources for user research: Case Habbo Hotel, *Proceedings of the 28th Conference on Information Systems Research* in Scandinavia (IRIS'28). http://www.soberit.hut.fi/johnson/Johnson_IRIS_2005.pdf

Johnson, M. (2007a). "Unscrambling the 'Average User' of Habbo Hotel." *Human Technology*, 3(2), 127–153.

Johnson, M. (2007b). *User developer dialogue in Habbo Hotel*. User Innovations Workshop, 11.5.2007 National Consumer Research Centre, Helsinki, Finland.

Leavis, F. R., & Thompson, D. (1933). *Culture and environment*. London: Chatto and Windus.

Lehdonvirta, V., Wilska, TA., & Johnson, M. (2009). Virtual consumerism: Case Habbo Hotel. *Information, Communication & Society*, 12(7).

Livingstone, S., Van Couvering, E., & Thumim, N. (2008). Converging traditions of research on media and information literacies: Disciplinary, critical, and methodological issues. In J. Corio, M. Knobel, C. Lankshear, & D. J. Leu (Eds.), *Handbook of research in new literacies*. Mahwah, NJ: Lawrence Erlbaum.

Masterman, L. (1980). *Teaching about television*. London: Macmillan.

Masterman, L. (1985). *Teaching the media*. London: Comedia.

Monroy-Hernández, A., & Resnick, M. (2008). Empowering kids to create and share programmable media. *Interactions*, 15(2), 50–53.

Resnick, M., Maloney, J., Monroy-Hernández, A., Rusk, N., Eastmond, E., Brennan, K.,...Kafai, Y. (2009). Scratch: programming for all. *Commun. ACM, 52*(11), 60–67.
Sulake website (n.d.). Retrieved on February 28, 2010, from http://sulake.com/habbo
Thompson, D. (1973). *Discrimination and popular culture* (2nd. ed.). Harmondsworth, UK: Penguin.

CHAPTER SIX

eHealth

Bridging the Divide between Current Performance and Legitimate Expectations in Health Care Delivery

Luca Camerini & Yujung Nam

The intersection between the power of networking technologies and the affordances exploited by individuals in order to address their information and communication needs has created a new societal configuration that is still largely unexplored. The healthcare domain is a substantive part of this ever-changing environment. On one hand, the Internet has empowered patients by providing them with information that they can bring to their physicians, allowing patients to be better informed. On the other hand, institutional actors such as research labs, clinics, and hospitals have become increasingly intertwined. As often happens with technology, the social impact of nascent fields, such as telemedicine and health informatics, has been largely underestimated. More systematic research is needed in order to deepen our understanding of the dynamics that structure the design, implementation, and evaluation of eHealth interventions. This need has given birth to the eHealth research field, which lies at the intersection of Internet research and health communication.

This chapter is a brief overview of the field of eHealth. It starts with a general description of the context and rationale behind this line of research, moves to a discussion of the field's theoretical foundations and ends with an overview of contemporary types of eHealth interventions. It is difficult to do justice to an entire field of research in a single chapter. Therefore, we will only present the milestones of the field, stressing points that could become potential research topics in the future. Throughout the review of the main theoretical and empirical findings, various issues will be raised that are at the heart of the field. Some of the points that we will discuss include the shift from an individual approach to the study of eHealth interventions to a social networking perspective. Along these lines, we will examine the lengthy path to construct theories specific to addressing eHealth issues and the modern tools and channels that can be exploited in order to enhance empirical verification and the interventions' effectiveness. We hope that these points will provoke critical discussion about the new frontiers of eHealth research in the future.

The Context of eHealth

The Rise of eHealth: Preliminary Definitions

The Internet is one of the most widely adopted communication technologies in the world today (Internet World Stats 2009). The communicative practices within the healthcare systems are no exception. According to the Pew Internet & American Life Project, more than 75 percent of American adults have used the Internet to find medical information (Pew Internet & American Life Report, 2008). Similar data can be seen amongst European countries (Internet World Stats, 2009). The Internet is not only used to locate health information, but also to locate treatment plans, buy drugs, and keep track of health data. The widespread use of the Internet for such a variety of purposes demonstrates the need for a systematic approach to the study of its impact on the healthcare system and the individuals who navigate it. For this reason, a new field of research, eHealth, has been created. Currently, one of the most widely recognized definitions of eHealth states that it is the use of emerging information and communication technology, especially the Internet, to improve or enable health and healthcare (Eng, 2001). According to Eysenbach (2005), this field has grown in the intersection of medical informatics, public health, and business and is defined as referring to health services and information delivered or enhanced through the Internet and related technologies. The scope of eHealth, according to these definitions, is very large and includes technologies that might not necessarily be web-based. A literature review conducted by Oh, Rizo, Enkin, and Jaded (2005) found more than 51 definitions for the term. They also found that despite the terminological disagreement a number of these definitions have common characteristics including the use of technology as the main communication channel and a focus on health-related issues. The combination of these two components results in eHealth interventions, which Bennett and Glasgow (2009) defined as "systematic treatment/prevention programs, usually addressing one or more determinants of health (frequent health behaviors), delivered largely via the Internet (although not necessarily web-based), and interfacing with an end user."

Online Health Information Seeking: Potential Reach and Users' Profiles

An increasing amount of data about the potential reach of eHealth interactive communication applications has been reported by various sources. A general finding of previous research on the subject shows that Internet access and usage is linked to socio-demographic factors. In the United States, 92 percent of adults aged 18–29 use the Internet, while only 37 percent of adults over 65 regularly look for information on the Internet (Pew Internet & American Life Report, 2008).

The speed of the typical Internet connection is increasing and net penetration is 21.9 percent worldwide, with the highest rates in the United States (73.6 percent), Europe (43.4 percent), and Asia (15.3 percent). In Europe, the highest numbers of Internet users are found in Germany, the United Kingdom, and France, whereas in Asia the most active countries are China, Japan, and India. The growth in Internet usage has been observed around the world, but most particularly in Africa and the Middle East (Internet World Stats, 2009), although penetration (especially of broadband) in these regions still lags far behind much of the rest of the world.

Not surprisingly, one of the most prominent reasons for individuals to go online is to locate health information. When ranking their web activities, individuals surveyed stated that seeking health information was fourth behind email communication, search engine utilization, and map browsing. When searching for health information, individuals tended to search for information on specific diseases (66 percent), medical treatments (55 percent), physical exercises (52 percent), and doctors and other health professionals (47 percent) (Pew Internet & American Life Project, 2009). Individuals affected by chronic conditions, such as cancer, diabetes, and AIDS/HIV represented the largest proportion of ePatients.

Despite the understandable enthusiasm that these statistics generated among eHealth developers and promoters, barriers still exist that might prevent non-Internet users from going online, such as unreliable infrastructure, especially in developing countries; a lack of the necessary skills and difficulty reading and/or understanding online content (Cullen, 2006; Norman & Skinner, 2006). While some of these barriers will likely be solved by technological development (i.e., more robust infrastructure), others, such as the quality of health information, are being addressed directly by social scientists.

The Benefits of eHealth Interventions

There are two ways to evaluate the effectiveness of eHealth interventions: the first is the potential reach of the interventions, and the second, the impact of the intervention on the patient's health status. Whether this impact is direct, mediated, or moderated will be discussed later. However, it is important to note that several randomized studies have shown the effectiveness of Internet interventions within a variety of clinical studies, including asthma management, smoking cessation, chronic pain, diabetes, alcohol abuse, anxiety and depression, headache, insomnia, HIV prevention, mental health disorders, physical activity, organ donation, sexually transmitted diseases, stress management, and weight loss (see Bennett & Glasgow [2009] for a review). Furthermore, within these clinical outcomes, medical decision-making and risk assessments have proven to be well-supported by interactive web applications.

Evidence of the effectiveness of eHealth interventions is well-documented in recent literature. For example, Griffith, Lindenmeyer, Powell, Lowe, and Thorogood (2006) conducted a qualitative systematic review in order to understand the reasons why healthcare interventions are delivered over the Internet. The main result of this study was the creation of a categorization of reasons for using the Internet, as well as drawbacks, and suggestions for future research. The six main reasons for using the Internet identified were: reducing costs for users, reducing costs for health services, reducing isolation of the user, the need for timely information, reducing the social stigma, and increased user and supplier control of the intervention. As for drawbacks, the authors found that the Internet might reinforce the problems interventions are intended to help and, therefore, the Internet should not become a substitute for face-to-face interactions (e.g., patients may overestimate the potential of a peer support group and become more reluctant to meet their physician). Additionally, the authors suggested incorporating the costs for the users and their social networks into the study, not limiting the scope of the healthcare system and comparing the Internet with traditional care in order to mark its added benefits.

It is important also to note that a revised version of the Cochrane review by Murray, Burns, See, Lai, and Nazareth (2005) revealed that interactive health communication applications (IHCA) have had a positive effect on knowledge, social support, behavioral outcomes, and clinical outcomes. IHCA are also likely to have a positive effect on self-efficacy. The impact of IHCA on emotional and economic outcomes was not able to be determined.

Internet interventions are not beneficial simply as a supplementary treatment option, but they have also been proven to be more effective than traditional interventions in various domains. In particular, Wantland, Portillo, Holzemer, Slaughter, McGhee (2004) reviewed clinical trials investigating the difference between web-based versus non-web-based interventions. Sixteen out of the seventeen studies included in the analysis favored the web-based intervention. However, only six of these studies showed a significant improvement when compared to their offline counterparts. This is a general problem of Internet-based randomized controlled trials, which often suffer from a high attrition rate as stated by Eysenbach (2005). However, eHealth intervention literature shows that regardless of these attrition rates, the benefits of eHealth interventions are consistent (Strecher, 2007; Van Meter, 1999; Ybarra & Eaton, 2005).

The Pitfalls of eHealth Interventions

Although shown in previous research to be of only mild impact, possible pitfalls do exist when implementing eHealth interventions. Kraut, Patterson, Lundmark, Kiesler, Mukopadhyay, and Scherlis (1998) conducted a major field experiment on the effects of Internet usage within a clustered population in 1998. They found that individuals communicating via the Internet were more likely to be socially isolated, lonely, and depressed. Although a follow-up study partially reversed these negative results (Kraut et al., 2003), other concerns about the unintended effects of Internet usage have been underlined in literature on virtual communities. Some authors have reported that the lack of visual cues in online interactions were one of the most relevant disadvantages (Brennan, 1996; Finn & Lavitt, 1994; Sharf, 1997). Another disadvantage is the high level of (new) media literacy required to access online communities, with literacy distributed according to the usual demographic stratifications. Additionally, despite the claim that the Internet can help reduce social disparities, when it comes to socio-demographics it is very likely that the types of individuals using these kinds of systems represent a less diverse group than assumed (Finfgeld, 2000).

Another relevant concern about the use of online communities is the possibility of generating an Internet addiction (Madara, 1997). Such addictions have two main damaging effects, namely psychological disorders and lag-time before face-to-face consultations. A blind trust in the virtual community's support can cause delays in seeking necessary medical face-to-face counseling (Finn, 1996). In addition, from a health communication perspective, there is much concern over the quality of the information exchanged within online communities—particularly within communities not moderated by health professionals (Eysenbach, Powell, Kuss, & Sa, 2002; Klemm, Reppert, & Visich, 1998.)

THEORETICAL INSIGHTS

Until recently, the need for scientific legitimization of eHealth research favored the delineation of a landscape populated by empirical works. The main focus of these qualitative and quantitative studies was on the effectiveness of eHealth applications. Several individual and social parameters have been investigated (i.e., self-efficacy, social support, empowerment, health literacy, etc.), with the ultimate focus on improvements in the patients' health outcomes. However, in order to gain a complete overview of the eHealth domain, it is important to examine two theoretical approaches that are used in eHealth research designs.

The first, and perhaps more classical approach, relies on the idea that the best predictor for health outcomes is a correct (i.e., healthy) individual behavior. We refer to it as the *individual approach* (IA). The main theories that belong to this approach are meant to explain behavior change as an outcome of different individual characteristics. In this scenario, eHealth interventions are usually considered to have a moderating effect on these individual characteristics. Typical investigations under this individual umbrella are concerned with how eHealth applications impact, among other things, a user's self-efficacy, knowledge, and intention to behave.

The second approach is more recent, and will likely gain importance in the scientific community of eHealth researchers in the near future. We refer to it as the *social networking approach* (SNA). As previously described, traditional approaches by social scientists and psychologists put emphasis on individualistic explanations in describing how social actors take actions. However, these approaches often lack nuanced understanding of the social contexts in which these interactions take place and how other social actors directly or indirectly influence the individual's actions. The SNA approach assumes that a social action takes place within a social system that involves many other actors whose perceptions, beliefs, and actions become social reference points in the individual's decision-making process (Knoke & Kuklinski, 1996). Different social phenomena and the adoption of specific behaviors, such as smoking, eating, and exercising, as promoted by public health interventions, spread through social connections within diverse network structures (Buller et al., 2000; Burt, 1987; Christakis & Fowler, 2007; Granovetter, 1973; Klovdahl, 1985; Valente, Hoffman, Ritt-Olson, Lichtman, & Johnson, 2003). As we are all embedded in social groups, the views and behaviors of close networks can influence our attitudes and tolerance for health-related choices and behaviors through social and psychological mechanisms.

IA and SNA, along with other theories, have attempted to explain and predict health behavior at multiple levels of analysis, most particularly at the intrapersonal, interpersonal, group, organizational, and community levels. Previous efforts to compare theories in terms of their efficacy in predicting health behaviors are well-established in the scientific literature and supported by empirical investigations. Nevertheless, the theories that they encompass were not originally developed to address eHealth interventions, but draw on preexisting psychological, sociological, and communication theories. Prioritizing the definition of eHealth-specific theories and models is a nexus for the theoretical development of eHealth research. Examples of such attempts are mostly tied to the individual approaches and can be found in Eysenbach et al. (2002), Murray, Burns, See, Lai, and Nazareth (2005), and Dutta-Bergman (2006). It is important to use and test theories in quantitative and qualitative research in the eHealth domain in order to improve, extend, and modify where appropriate, and it is hoped that these efforts will be extended in the future.

We will now briefly review the main theories that belong to the individual and social networking approaches.

Individual Approaches to Behavior Change

Self-efficacy theory (SET)

SET was developed from the social cognitive theory developed by Albert Bandura (1977). Although not the first model to appear in the literature, it is certainly the one that has had the greatest impact in regard to explaining behavior change. According to SET, behavior is determined by the expectations of an individual with respect to two variables: self-efficacy and expected outcome (Bandura, 1977). Self-efficacy is an individual's belief about their abilities to carry out or engage in a behavior and achieve the desired outcome. The expected outcome is what they are convinced will occur after adopting a certain behavior. According to Bandura, self-efficacy is the most important factor in the behavior change process.

Theory of reasoned action (TRA)

The TRA was developed by Fishbein and Ajzen and states that human behavior is determined by the intention of the individual to engage or not engage in a certain behavior (Fishbein & Ajzen, 1975). Therefore, the intention of an individual is dependent upon the attitude they have toward a certain behavior as well as their perception of how other people evaluate this behavior. On the one hand, the attitude is regarded as the result of beliefs or expectations that the individual has over the final outcome of a certain behavior. On the other hand, the motivation to trust other people's discussions generates a series of subjective norms that influence the decision in relation to the behavior.

Theory of planned behavior (TPB)

The TPB is an extension of the TRA and was introduced by Ajzen (1988). The basic assumptions are the same as in the TRA, but Ajzen notes that the intention to undertake a certain behavior cannot be the only factor that determines a successful decision-making process, especially in cases where the individual has no control over the behavior itself. To address this problem, Ajzen introduced the concept of perceived behavioral control, which is the individual's perception in regard to the difficulty of accomplishing a certain behavior. This new concept, in turn, depends upon control belief, which states that the person holds some beliefs about the factors that may facilitate or hinder the behavior. This concept is closely tied to the idea of self-efficacy. The TPB has proven more effective than the TRA in predicting the reactions of the individual behaviors connected to an individual's health.

Transtheoretical model (TTM)

The TTM was developed by a team of researchers under the guidance of Prochaska and DiClemente (1983). Unlike previous models, the TTM focuses on the process of behavior change rather than on its determinants. According to these authors, the individual begins to take some action by passing through six developmental stages: pre-contemplation, contemplation, preparation, action, maintenance, and termination. In the pre-contemplation stage, the person has the attitudes and information that favor behavior change. In the contemplative stage, the person starts considering and expressing a willingness to change. The preparation stage immediately precedes the action itself. Action is the stage where a behavioral change becomes observable. The maintenance phase includes the time frame during which the person participates in the new behavior. Finally, the termination phase is observed when the person persists in adopting a certain behavior based solely on his awareness and control over his actions, thus gaining a high level of self-efficacy. The transition from one phase to another occurs through various processes, such as awareness and counter-conditioning, which are not specifically addressed here (see Prochaska & DiClemente [1983] for further details).

TTM can be used to identify where an individual is with respect to their general knowledge about an action and its risk factors and can be used by others to advance the individual to the next stage in the process via health campaign or intervention message.

Stage theories have three principal elements. The first element is a category system or prototype for each stage. However, the boundaries between each stage may be unclear and need clarification depending upon the situation. The second element is an ordering of the stages, as individuals are not expected to move along the stages in a linear fashion or at the same rate. The third element is factors that prevent advancement or movement at each stage. These are often identified as barriers that inhibit the individual from advancing to higher stages.

Elaboration likelihood model (ELM)

The ELM was developed by Petty and Cacioppo (Petty & Cacioppo, 1986). Much like the TTM, this model describes the process of change at the individual level. The main assumption of the ELM is that people constantly process the information that they encounter. This information processing can be more or less conscious and rational. According to the authors, two routes exist through which individuals process information and act: the central and peripheral routes. Central route processing involves a deep scrutiny of the arguments included in the messages. The person evaluates the content and carefully weights the decision to act. Alternatively, peripheral route processing relies on superficial aspects of messages, such as the perceived credibility of the source, format, and attractiveness of the slogan. More specifically, two variables can predict which route the individual is more likely to take: the motivation of the person in relation to the contents of the message and his ability to critically evaluate the arguments. These two variables are often associated with the concept of individual involvement in that the more a person is involved in the topic at stake, the more likely he will centrally process the message.

Health belief model (HBM)

The HBM was introduced by Rosenstock and developed further by other researchers (Janz & Becker, 1984; Mattson, 1999; Rosenstock, 1974). This theory is based on the interaction of six components: severity, susceptibility, benefits, barriers, incentives to action, and self-efficacy. The first two concepts, severity and susceptibility, are complements of each other. The level of severity is the result of the perceptions that the person has in regard to the possible outcomes of a certain behavior. Susceptibility refers to what the person is willing to accept as a possible negative outcome of the behavior. Both of these determinants must be developed so that the person can consider whether to engage in or change a certain behavior. It is at this decision-making point that the concept of a benefit is important. A benefit is the individual's belief in regard to the effectiveness of possible countermeasures that might mitigate the negative outcomes of the behavior. Finally, self-efficacy concerns the perception of the individual in relation to his actual capacity to adopt the behavior that would give him the above benefit.

Most IA theories in widespread use, such as the TRA or the HBM, have, as a central component, a balance between what an individual expects to be the cost of engaging in a health behavior versus what is perceived to be the benefit. Most theories seek to identify the beliefs and other factors that influence this equation in the individual's mind. Researchers attempt to create regression equations based on factors that will yield the relative probability of performing the health behavior for the individual in question. In intervention or prediction of eHealth studies, individuals are usually arrayed along an action continuum from those most to least likely to perform the desired behavior or to deter an undesired risk behavior. Intervention studies, therefore, aim at moving people to a higher point along the continuum and increasing the likelihood of the desired health behavior.

Social Networking Approaches to Behavior Change

Social Capital

Social capital is a crucial feature of social organization that regulates and guides interactions in group contexts and makes cooperation for mutual benefit possible. Such cooperation is integral to studies of group support and individual interactions within a community (Putnam, 1995; Putnam, Leonardi, & Nanetti, 1993). The term "social capital" was first used to describe the "intangible substances" that account for a community working together toward a common goal (Hanifan, 1920), but received significant attention after Bourdieu (2001) and Coleman (1988) made major contributions. Bourdieu treated social capital as a narrow measure of the benefit of membership in clearly established organizations, whereas Coleman's research provided a more encompassing, broader definition. Most recently, Putnam conducted a number of in-depth studies aimed at providing a clearer understanding of the impact of changes in social capital on a community, with social capital defined as "features of social organization, such as networks, norms, and trust that facilitate coordination and cooperation for mutual benefit. Social capital enhances the benefits of investment in physical and human capital" (Putnam, Leonardi, & Nanetti, 1993, p. 35).

As a general concept, social capital is quite broad in scope and encompasses a multitude of social interactions. A distinction is often made between two types of social capital. Bridging social capital refers to inclusive interactions that draw people together across social boundaries, whereas bonding social capital refers to reinforcing interactions that reaffirm exclusionary relationships (Putnam, 1995; Wuthnow, 2002). Bridging social capital is an important component of social interactions because it is through bridging interactions that an individual is able to tap into heterogeneous groups and access unique resources. A strong cache of bridging social capital is important as it has the potential to connect an individual to a unique, previously untapped set of resources (Leonard & Onyx, 2003). Bonding social capital is important because it provides individuals with emotional and social support (Williams, 2006) and is a measure of how strongly an individual is connected with his immediate support network. Despite some disagreement within the field over how social capital is measured (Hawe & Shiell, 2000; Woolcock & Narayan, 2000), most scholars agree that social capital has a strong and direct influence on social interaction (Putnam, 1995; Putnam, Leonardi, & Nanetti, 1993). Notably, Putnam's comprehensive survey of social capital in America found that a strong correlation existed in regard to the fluctuations in social capital and indicators of general health.

Social capital has been established as a theoretical frame and key component of health outcomes, not only through an individual's perception that a close social support network is available, but also through having recourse to bridge social capital, as the ties created by "friends of friends" provide exposure to new health contacts, information, and resources. Kawachi, Kennedy, and Glass (1999) reported that after individual-level factors, such as low income, fewer years of schooling, and smoking, were held constant, low social capital was still a major factor in self-rated poor health. As social networking becomes increasingly popular across different generations, including those only able to connect to networks occasionally through access at public sites or at work, a new wellness approach should be used, in which a networking and information-sharing model is used to disseminate health information and build social capital through connections to others with cognate health concerns.

Diffusion of innovation

Diffusion of innovations theory attempts to explain how new ideas and practices spread in communities. It shows that various social (rather than economic) factors facilitate or accelerate the diffusion of innovations. The premise of this theory is that new ideas and practices are disseminated through interpersonal contacts largely consisting of interpersonal communication (Beal & Bohlen, 1955; Katz, Downs, Cash, & Grotz, 1970; Rogers, 2003; Ryan & Gross, 1943; Valente, 1996). Interpersonal influence has been consistently shown to accelerate the diffusion process, particularly if a degree of risk or uncertainty is attached to it. Rapid diffusion of social behavior is mostly accompanied by the approval of the innovation by influential persons, known as opinion leaders, within social networks or social sectors. New ideas and practices spread within and between communities when these intermediaries have a favorable view of the innovation. However, individuals in equivalent roles or positions in a hierarchy who are exposed to the same ideas and concepts can also spread the innovation, as can mass media outlets depicting positive consequences for the adopters. As such, diffusion approaches have helped to accelerate the adoption of best practices in various social programs by effectively identifying intermediaries who have the greatest influence over behavioral influence. Diffusion research examines professional associations, sophisticated networks, and technology for potential intermediaries to target for adoption.

EMPIRICAL INSIGHTS: UNPACKING THE BLACK BOX

In the vast majority of eHealth experiments, actual intervention is considered the "black box." Studies on interactivity (Rafaeli, 1988) stressed the need for a more refined definition of the enabling functions of Internet technology. Also, consider that Bennett and Glasgow (2009) restate the need for specificity when referring to interactivity in eHealth:

> Although there is emerging empirical support for Internet interventions, questioning whether the Internet works as a platform for intervention delivery may have inherent flaws. Doing so [...] belies the importance of specific intervention designs and components. Indeed, the approach of collapsing across extremely heterogeneous interventions and study outcomes (whether conceptually or analytically) is problematic because it masks important variations that may be necessary to understand how to improve intervention effectiveness (Eysenbach, 2005). At a minimum, reviews and meta-analyses should focus on interventions for specific outcomes. Optimally, however, we will begin to see more factorial study designs testing the utility of varying intervention components (p. 277).

In the last few years, the digital revolution has changed the landscape of both interpersonal and mass communication and, as such, much of communication is now mediated by personalized, immediate, and more persuasive forms, such as RSS feeds, mobile phone conversations, text messages, targeted e-mails, instant messages, social networking services, and personal web blogs. The number of choices for communication modes and devices for health consumers has grown exponentially and the traditional boundaries and categorizations in health interventions are blurring.

As such, in order to actualize and exploit the potential reach of interactive eHealth applications, health communicators must understand this new media landscape and the types of components or "tools" found in such interventions in order to take advantage of opportunities for greater and more effective areas of reach and to understand the challenges presented by a new and bewildering media landscape. We, therefore, propose a typology of these tools, which functions both as a review and an attempt to systematically address the new frontiers of Internet technology in the healthcare domain.

Consumers are embracing the enormous changes in their media choice options, which are enabling them to conveniently construct and sift through personalized input and customized content in digital media. Such abilities indicate that print and broadcast news media will be increasingly less effective in health campaigns and dissemination initiatives. The remarkable growth of RSS feeds, personal web pages, targeted web portals, special purpose social network web vehicles, TiVo, blogs, podcasting, and customized cell phone ringtones epitomize consumer needs for personally relevant and customized media content.

This trend reveals that active and informed consumers are taking charge of and fulfilling their communication needs. Passive acceptance of information is increasingly becoming obsolete. The uses and gratifications provide context for the active participation of health consumers. Individuals choose to meet their needs through mediated channels ((Katz, E., Blumler, J. G., & Gurevitch, M., 1974; Rubin & Windahl, 1982; Ruggiero, 2000). New media have created new sets of motivation, and needs among audience groups. Lin (1996; , 2002) identified the new dimensions that have been created from varying situations, media outlets, and individual situations along the following grid: relaxation, companionship, habit, passing time, entertainment, social interaction, information/surveillance, arousal, and escape. These communicative needs interact with other variables, such as costs, demographics, personal differences, and personal preferences of different media.

We loosely categorized interventions into four categories, which are then matched with examples of content, digital media devices used to fulfill these desires and needs in the field of eHealth (Figure 1).

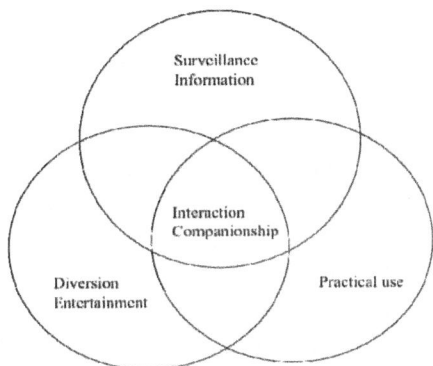

Figure 1. eHealth communication needs

Surveillance and Information Interventions

Fulfilling the need to gain important knowledge in order to accomplish goals. This may mean more targeted information for decision-making related to prognosis or diagnosis of health conditions.

Informational needs might trigger health consumers to search web services such as WebMD or the Mayo Clinic, or rely on specific search results from Google, Yahoo or other search engines. Static online content, such as individual homepages, blogs, web forums, and chat rooms may serve informational purposes.

Norman, Zabinski, Adams, et al. (2007) concluded in their systematic review of eHealth intervention programs between 2000 and 2005 that the most commonly used eHealth components were a combination of websites and e-mail followed by websites alone. Studies have shown that the integration of healthcare systems and information through easily accessible web portals, such as WebMD and the Mayo Clinic, is crucial in overcoming time and space limitations in the patient-provider relationship.

Diversion and Entertainment Interventions

Fulfilling the need to be entertained, relaxed, and diverted from distressing physical and mental situations. This may mean idling and passing time or seeking arousal or escape.

In order to accomplish physiological and psychological health, we seek means to be entertained, relaxed, and diverted from issues. A wide range of sources exists to fulfill these needs. For health promoters and intervention designers, applications using strategies such as humor, sexual appeal, and other attention-getting devices have garnered attention as being useful to future health initiatives. The impact of Farrah Fawcett's colon cancer diagnosis and the broadcast of her struggle on a popular TV show may have had an important mediating variable in health communication. At the same time, patients may find mediated amusements and diversions by playing customized video games that contain personalized health messages and information necessary to improve their attitudes and beliefs. Intervention studies have shown that a health intervention that is more appealing and entertaining can achieve more than less appealing or entertaining interventions (Atkinson & Gold, 2002; Baranowski et al., 2003). When websites are designed as engaging and contain dynamic multimedia games, intervention programs achieved higher completion rates and led to positive findings in general (Baranowski et al., 2003). Some interactive computerized interventions utilize computer-tailored materials combined with personalized video games and dynamic online content. Interactive games and dynamic web content often use innovative functions, such as haptics and location-based knowledge presentations and tracking, in order to accurately assess and tailor intervention effort feedback (Dinse et al., 2005). Potential advantages for designing behavioral interventions through immersive, tailored computer games include enhanced levels of cultural sensitivity (Atkinson & Gold, 2002; Nigg, 2005) and easy access to self-monitoring processes with specific feedback. For example, participants in a virtual weight loss program filled out virtual logs for exercise and food consumption (Tate, Wing, & Winett, 2001; Vandelanotte et al., 2005). Health professionals responded with personalized feedback according to the given calorie intake and energy expenditure provided by the users. Both studies showed positive results in regard to the interactive weblog combined with timely and personalized feedback. Such results reaffirm the benefits of eHealth interventions on behavioral changes and patient empowerment.

Practical Use Interventions

Direct applications of digital devices for diagnosis, treatment, or communication between providers and patients. Means of purchasing health-related products, such as drugs or medical equipment.

Consumers need to access various web-based services in order to purchase drugs and other medical equipment. Online drug stores and web portals that streamline users' economic desires for purchasing health-related goods and services have increased. Intervention programs have targeted influencing dietary behavior by changing the shopping habits of participants by using interactive web technology, such as placing informational kiosks in grocery stores (Andersonet al., 2001). eHealth designers need to test various components that address such needs at multiple levels, such as through websites, software programs, multimedia components, e-mail, e-newsletters, interactive phone technology, or interactive kiosk systems.

From the healthcare providers' perspective, unobtrusive, inexpensive, and mobile devices for measuring patients' physiological conditions related to health issues or digital communication devices can assist patient-provider communication. For example, mobile devices for remote healthcare services, such as personal glucose monitoring devices or heart rate monitors, need to be deployed and evaluated in today's society more than ever (Tate, Wing, & Winett, 2001; Womble et al., 2004; Yoo et al., 2003). Video conference devices and technology used in the field of psychiatry and e-counseling have improved the outcomes of several intervention studies (Pinto & Ricardo, 2005).

Socio-cultural Interaction and Companionship

Fulfilling the need to interact, socialize, support, and validate with others for various purposes.

Communication needs are the basis of the usage of every media, and are extremely important in the health context. For example, registering at PatientsLikeMe.com and belonging to a support group for a specific condition is understood to be an interaction behavior. Countless YouTube clips exist that are personal testimonials or are related to health and medical procedures. People share personal experiences, highly specialized knowledge, and perspectives on certain conditions. Trust and support may serve beyond the customary medical boundaries and may provide spiritual support from some as well. Gaining self-awareness and spiritual support from others in a trusted environment can have a positive impact on medical outcomes.

Social networking vehicles such as Facebook and MySpace host countless groups and fan pages related to medical issues and conditions. GPS-based services have gained recognition and popularity among consumers and made physical interaction easier by providing users with the ability to locate each another or information sources near them. Health institutions and organizations creating public service announcements have increasingly tailored their messages to mobile and text message formats in order to more quickly disseminate information.

Studies have found higher participation and engagement in Internet programs that have included peer support groups (McKay et al., 2002). The LIFE Community project used a mobile social networking approach in order to assist successful cancer survivorship for young adults and attempted to build health social capital, enable easy access to useful resources, and increase awareness of long-term side effects of treatment, environmental hazards and stressors, and options in adult health care management (McLaughlin et al., 2010). Participants were issued a pre-paid cell phone with SMS and MMS capabilities, as well as an integrated video camera capable of taking short low-resolution videos. They received periodic SMS messages describing visual narrative assignments such as "introduce yourself to the group" or "teach us about your experience as a cancer survivor." Their direct speech or interviews with significant others were posted on the LIFE Community social networking website and shared with other cancer survivors. The SMS and MMS messages, comments, in-site email service, and friend functions were similar to commercial social networking web services, such as Facebook, and provided seamless communication among the participants.

The preliminary results of this study show that when combined with face-to-face meetings and phone calls, social-networking solution-based interventions outperformed in-person only workgroups. Factors such as depression, perceived social support, general evaluation of intervention programs, information needs, information-seeking habits, and the level of self-efficacy interacted with the results. These results emphasize the significance of a participant-oriented approach in the way the users interact with the intervention technology (Ogden et al., 2006).

It should be noted that users' needs can overlap and be important in different ways to different people. Asynchronous communication through the Internet now allows individuals to exchange information, provide mutual support and search for services at their convenience simultaneously (McKenna & Bargh, 2000).

Searching for health related information may be a diversion to some patients, and connecting to other patients may have informational significance to others. Each health consumer's communication needs may vary based on their circumstances and demographics, as well as cultural and other personal variables. Therefore, these needs will be expressed in diverse ways through utilizing different digital media devices and contents. Cohen and Kaczorowski (2004) reported differences between life stage groups in their health communication. It seems clear that health communicators must be aware and sensitive to variations in media usage by different segments of the population when they design and evaluate the digital media content usage and devices to be used to disseminate health content aimed at assisting positive behavior changes in patients or the general public.

Conclusion: A Nexus from the Lessons Learned

Research on the impact of eHealth interventions is no longer in its infancy. Throughout this chapter, we have shown how this field has been able to build upon theories and applications borrowed from different disciplines. The problem of exploiting networking technology in order to reduce the divide between consumers' expectations and healthcare systems' performances may remain unsolved. However, the constant effort to integrate eHealth applications into healthcare delivery is proving beneficial in terms of social support, empowerment, health literacy, and health outcomes. This scientific enquiry is far from complete. Our overview suggests future research opportunities or nexuses, as summarized later in the chapter, with distinctions kept between those linked to the theoretical dimension of the discipline and empirical research.

The two main theoretical dimensions that emerge from this research are the integration of the social networking approach in the field of eHealth and Health communication, and the need for eHealth-specific theories and models. Individual and networking approaches to the study of eHealth interventions focus on different aspects of their impact, and should be considered as complementary rather than oppositional. Indeed, Internet interventions in healthcare impact characteristics, such as knowledge, empowerment, motivation, and intention to behave. These characteristics are strongly affected by the social network in which the individual is embedded and, thus, the concepts of social support, social capital, diffusion and adoption become extremely relevant in regard to gaining a comprehensive understanding of what does and does not work in terms of technological support for healthcare delivery. Individual approaches are grounded in the tradition of psychology and, more recently, social marketing. The networking approaches have been developed from a sociological and media studies perspective. These contributions are merging in the eHealth and health communication domains with results that are promising and likely to be refined in the near future. This integrative perspective requires the definition and empirical verification of theories and models that are specifically aimed at exploring eHealth interventions. As we underlined previously, these kinds of theoretical contributions are still scarce in the eHealth domain. After almost ten years of qualitative and quantitative studies on the impact of Internet technology in healthcare, the need for further efforts to enrich the theoretical dimension of the discipline persists.

From an empirical perspective, we mentioned a series of interventions that responded to the current uses and needs for gratification of the individual accessing eHealth tools. However, at least three challenges still need addressing: namely, the evaluation of single components that constitute an eHealth intervention, the need for an integration of new web technologies into a single intervention, and some level of agreement in respect to the metrics for eHealth evaluation. The attempt to "unpack" the black box that was introduced in this chapter underlies the fact that we must not limit experimental research in regard to the evaluation of eHealth interventions as a whole. Instead, an urgent need exists for more refined factorial designs in order to evaluate the effectiveness of the single tools that are offered to patients. This methodological challenge is likely to become increasingly relevant, considering the variety of tools that can be exploited, from podcasting to real-time monitoring. Eventually, the ways in which individuals use these tools will make a difference in terms of their effectiveness. However, in order to include actual usage in empirical models, it is necessary to reach an agreement on the metrics that are used to assess it (e.g., a consensus on the measurement units of weblogs metrics).

These theoretical and empirical challenges represent a nexus for the eHealth domain, grounded in the work done so far. The pursuit is ongoing, and future research on the impact of the Internet in the healthcare domain should aim to generate a solid point of reference for researchers, health professionals, and public health policymakers alike.

REFERENCES

Ajzen, I. (1988). *Attitudes, personality, and behavior* (U.S. edition ed.). Chicago, IL: Dorsey Press.

Anderson, E. S., Winett, R. A., Wojcik, J. R., Winett, S. G., & Bowden, T. (2001). A computerized social cognitive intervention for nutrition behavior: direct and mediated effects on fat, fiber, fruits, and vegetables, self-efficacy, and outcome expectations among food shoppers. *Annals of Behavioral Medicine, 23*(2), 88–100.

Atkinson, N. L., & Gold, R. S. (2002). The promise and challenge of eHealth interventions. *American Journal of Health Behaviour, 26*(6), 494–503.

Bandura, A. (1977). *Social Learning Theory*. Englewood Cliffs, NJ: Prentice Hall.

Baranowski, T., Baranowski, J. C., Cullen, K. W., Thompson, D. I., Nicklas, T., Zakeri, I. F., et al. (2003). The fun, food, and fitness project (FFFP): The Baylor GEMS pilot study. *Ethnicity and Disease, 13*(1; SUPP/1), 1–30.

Beal, G. M., & Bohlen, J. M. (1955). How farm people accept new ideas. *Cooperative extension service report, 15*.

Bennett, G. G., & Glasgow, R. E. (2009). The Delivery of Public Health Interventions via the Internet: Actualizing Their Potential. *Annual Review of Public Health, 30*, 273–292.

Bourdieu, P. (2001). The forms of capital. *The Sociology of Economic Life*, 96–111.

Brennan, P. F. (1996). The future of clinical communication in an electronic environment. *Holistic Nursing Practice, 11*(1), 97.

Buller, D., Buller, M. K., Larkey, L., Sennott-Miller, L., Taren, D., Aickin, M., et al. (2000). Implementing a 5-a-day peer health educator program for public sector labor and trades employees. *Health Education & Behavior, 27*(2), 232.

Burt, R. S. (1987). Social contagion and innovation: Cohesion versus structural equivalence. *American Journal of Sociology, 92*(6), 1287.

Christakis, N. A., & Fowler, J. H. (2007). The spread of obesity in a large social network over 32 years. *New England Journal of Medicine, 357*(4), 370.

Coleman, J. S. (1988). Social capital in the creation of human capital. *American Journal of Sociology, 94*(S1), 95.

Cullen, R. (2006). *Health information on the internet: a study of providers, quality, and users*: Greenwood Publishing Group.

Dinse, H. R., Kalisch, T., Ragert, P., Pleger, B., Schwenkreis, P., & Tegenthoff, M. (2005). Improving human haptic performance in normal and impaired human populations through unattended activation-based learning. *ACM Transactions on Applied Perception (TAP), 2*(2), 88.

Dutta-Bergman, M. J. (2006). A formative approach to strategic message targeting through soap operas: Using selective processing theories. *Health Communication, 19*(1), 11–18.

Eng, T. (2001). *The eHealth landscape: A terrain map of emerging information and communication Technologies in Health and Health Care*. Princeton, NJ: The Robert Wood Johnson Foundation.

Eysenbach, G. (2005). The law of attrition. *Journal of Medical Internet Research, 7*(1).

Eysenbach, G., Powell, J., Kuss, O., & Sa, E. R. (2002). Empirical studies assessing the quality of health information for consumers on the world wide web: a systematic review. *Jama, 287*(20), 2691.

Finfgeld, D. L. (2000). Therapeutic groups online: The good, the bad, and the unknown. *Issues in Mental Health Nursing, 21*(3), 241–255.

Finn, J. (1996). Computer-Based Self-Help Groups. *Computers in Human Services, 13*(1), 21–41.

Finn, J., & Lavitt, M. (1994). Computer-based self-help groups for sexual abuse survivors. *Social Work with Groups, 17*(1), 21–46.

Fishbein, M., & Ajzen, I. (1975). *Belief, attitude, intention and behavior: An introduction to theory and research*. Reading, MA: Addison-Wesley.

Granovetter, M. S. (1973). The strength of weak ties. *American journal of sociology, 78*(6), 1360.

Griffiths, F., Lindenmeyer, A., Powell, J., Lowe, P., & Thorogood, M. (2006). Why are health care interventions delivered over the internet? A systematic review of the published literature. *Journal of Medical Internet Research, 8*(2).

Hanifan, L. J. (1920). *The community center*. Boston: Silver Burdett.

Hawe, P., & Shiell, A. (2000). Social capital and health promotion: A review. *Social Science & Medicine, 51*(6), 871–885.

Internet World Stats. (2009). *Internet World Stats News* (No. 038): Internet World Stats.

Janz, N. K., & Becker, M. H. (1984). The health belief model: A decade later. *Health Education & Behavior, 11*(1), 1.

Katz, E., Blumler, J. G., & Gurevitch, M. (1974). Utilization of mass communication by the individual. In Blumler, J. G. & Katz, E. (Eds.), *The uses of mass communications: Current perspectives on gratifications research* (pp. 19–32). Beverly Hills: Sage.

Katz, S., Downs, T. D., Cash, H. R., & Grotz, R. C. (1970). Progress in development of the index of ADL. *Gerontologist, 10*(1), 20–30.

Kawachi, I., Kennedy, B. P., & Glass, R. (1999). Social capital and self-rated health: a contextual analysis. *American Journal of Public Health, 89*(8), 1187.

Klemm, P., Reppert, K., & Visich, L. (1998). A nontraditional cancer support group. The Internet. *Computers in nursing, 16*(1), 31.

Klovdahl, A. S. (1985). Social networks and the spread of infectious diseases: The AIDS example. *Social Science & Medicine (1982), 21*(11), 1203.

Knoke, D., & Kuklinski, J. H. (1996). *Network analysis*: Sage.

Kraut, R., Kiesler, S., Boneva, B., Cummings, J., Helgeson, V., & Crawford, A. (2003). Internet paradox revisited. *The Wired Homestead: An Mit Press Sourcebook on the Internet and the Family, 58*, 347.

Kraut, R., Patterson, M., Lundmark, V., Kiesler, S., Mukopadhyay, T., & Scherlis, W. (1998). Internet paradox: A social technology that reduces social involvement and psychological well-being? *American Psychologist, 53*, 1017–1031.

Leonard, R., & Onyx, J. (2003). Networking through loose and strong ties: an Australian qualitative study. *Voluntas: International Journal of Voluntary and Nonprofit Organizations, 14*(2), 189–203.

Lin, C. A. (1996). Standpoint: Looking back: The contribution of Blumler and Katz's uses of mass communication to communication research. *Journal of Broadcasting & Electronic Media, 40*(4), 574–581.

Lin, C. A. (2002). Perceived gratifications of online media service use among potential users. *Telematics and Informatics, 19*(1), 3–19.

Madara, E. J. (1997). The mutual-aid self-help online revolution. *Social Policy, 27*, 20–26.

Mattson, M. (1999). Toward a reconceptualization of communication cues to action in the health belief model: HIV test counseling. *Communication Monographs, 66*(3), 240–265.

McKay, H. G., Glasgow, R. E., Feil, E. G., Boles, S. M., & Barrera, M. (2002). Internet-based diabetes self-management and support: Initial outcomes from the diabetes network project. *Rehabilitation Psychology, 47*(1), 31–48.

McKenna, K. Y. A., & Bargh, J. A. (2000). Plan 9 from cyberspace: The implications of the Internet for personality and social psychology. *Personality and Social Psychology Review, 4*(1), 57.

McLaughlin, M., Gould, J., Nam, Y., Sanders, S., Qi, E., Meeske, K., et al. (2010). *A Mobile Social Networking Approach to a Cancer Survivorship Intervention for Young Adults*. Paper presented at the International World Congress of Psycho-Oncology Quebec, Canada.

Murray, E., Burns, J., See, T. S., Lai, R., & Nazareth, I. (2005). Interactive Health Communication Applications for people with chronic disease. *Cochrane database of systematic reviews (Online)*(4).

Nigg, J. T. (2005). Neuropsychologic theory and findings in attention-deficit/hyperactivity disorder: The state of the field and salient challenges for the coming decade. *Biological Psychiatry, 57*(11), 1424–1435.

Norman, C. D., & Skinner, H. A. (2006). eHealth literacy: Essential skills for consumer health in a networked world. *Journal of Medical Internet Research, 8*(2).

Norman, G. J., Zabinski, M. F., Adams, M. A., Rosenberg, D. E., Yaroch, A. L., & Atienza, A. A. (2007). A review of eHealth interventions for physical activity and dietary behavior change. *American Journal of Preventive Medicine, 33*(4), 336–345.

Ogden, C. L., Carroll, M. D., Curtin, L. R., McDowell, M. A., Tabak, C. J., & Flegal, K. M. (2006). Prevalence of overweight and obesity in the United States, 1999–2004. *JAMA, 295*(13), 1549.

Oh, H., Rizo, C., Enkin, M., & Jadad, A. (2005). What is eHealth (3): A systematic review of published definitions. *Journal of Medical Internet Research, 7*(1).

Petty, R. E., & Cacioppo, J. T. (1986). *Communication and persuasion: Central and peripheral routes to attitude change*: Springer.

Pew Internet & American Life Report. (2008). *The Engaged E-patient Population*: Pew Internet & American Life Report.

Pinto, A., & Ricardo, M. (2005). *Multiple video channel transmission using secure IP multicast*. ConfTele 2005, 5th Conference on Telecommunications, Tomar, Portugal.

Prochaska, J. O., & DiClemente, C. C. (1983). Stages and processes of self-change of smoking: Toward an integrative model of change. *Journal of Consulting and Clinical Psychology, 51*(3), 390–395.

Putnam, R. D. (1995). Bowling alone: America's declining social capital. *Journal of democracy, 6*, 65–65.

Putnam, R. D., Leonardi, R., & Nanetti, R. Y. (1993). *Making democracy work*: Princeton Univ.

Rafaeli, S. (1988). Interactivity: From new media to communication. *Sage Annual Review of Communication Research: Advancing Communication Science, 16*, 110–134.

Rogers, E. M. (2003). *Diffusion of innovations*. NewYork: FreePress.

Rosenstock, I. M. (1974). The health belief model and preventive health behavior. *Health Education Monographs, 2*(4), 355–385.

Rubin, A. M., & Windahl, S. (1982). *Mass media uses and dependency: A social systems approach to uses and gratifications*. Paper presented to the meeting of the International Communication Association, Boston, MA.

Ruggiero, T. E. (2000). Uses and gratifications theory in the 21st century. *Mass Communication & Society, 3*(1), 3–37.

Ryan, B., & Gross, N. C. (1943). The diffusion of hybrid seed corn in two Iowa communities. *Rural Sociology, 8*(1), 15–24.

Sharf, B. F. (1997). Communicating breast cancer on-line: Support and empowerment on the Internet. *Women & Health, 26*(1), 65–84.

Strecher, V. (2007). Internet methods for delivering behavioral and health-related interventions (eHealth).

Tate, D. F., Wing, R. R., & Winett, R. A. (2001). Using Internet technology to deliver a behavioral weight loss program. *JAMA, 285*(9), 1172.

Valente, T. W. (1996). Social network thresholds in the diffusion of innovations. *Social Networks, 18*(1), 69–89.

Valente, T. W., Hoffman, B. R., Ritt-Olson, A., Lichtman, K., & Johnson, C. A. (2003). Effects of a social-network method for group assignment strategies on peer-led tobacco prevention programs in schools. *American Journal of Public Health, 93*(11), 1837.

Van Meter, K. M. (1999). Social capital research literature: Analysis of keyword content structure and the comparative contribution of author names. *Connections, 22*(1), 62–84.

Vandelanotte, C., De Bourdeaudhuij, I., Sallis, J. F., Spittaels, H., & Brug, J. (2005). Efficacy of sequential or simultaneous interactive computer-tailored interventions for increasing physical activity and decreasing fat intake. *Annals of Behavioral Medicine, 29*(2), 138–146.

Wantland, D. J., Portillo, C. J., Holzemer, W. L., Slaughter, R., & McGhee, E. M. (2004). The effectiveness of Web-based vs. non-Web-based interventions: A meta-analysis of behavioral change outcomes. *Journal of Medical Internet Research, 6*(4).

Williams, D. (2006). On and off the 'net: Scales for social capital in an online era. *Journal of Computer Mediated Communication-Electronic Edition, 11*(2), 593.

Womble, L. G., Wadden, T. A., McGuckin, B. G., Sargent, S. L., Rothman, R. A., & Krauthamer-Ewing, E. S. (2004). A randomized controlled trial of a commercial internet weight loss program. *Obesity, 12*(6), 1011–1018.

Woolcock, M., & Narayan, D. (2000). Social capital: implications for development theory, research, and policy. *The World Bank Research Observer, 15*(2), 225.

Wuthnow, R. (2002). Religious involvement and status-bridging social capital. *Journal for the Scientific Study of Religion, 41*(4), 669–684.

Ybarra, M. L., & Eaton, W. W. (2005). Internet-based mental health interventions. *Mental Health Services Research, 7*(2), 75–87.

Yoo, S. K., Park, I. C., Kim, S. H., Jo, J. H., Chun, H. J., Jung, S. M., et al. (2003). Evaluation of two mobile telemedicine systems in the emergency room. *Journal of telemedicine and telecare, 9*(Supplement 2), 82.

CHAPTER SEVEN

Fielding Networked Marketing

Technology and Authenticity in the Monetization of Malaysian Blogs

JULIAN HOPKINS & NEAL THOMAS

This chapter draws upon extensive fieldwork[1] to analyse the relatively recent phenomenon of Malaysian "Lifestyle" bloggers writing paid "advertorials," and looks at potential consequences in terms of the underlying practices and values of blogging. Personal bloggers, from which the Lifestyle genre emanates, are a category underrepresented in academic studies, despite constituting the majority of bloggers (Cenite, Detenber, Koh, Lim, and Ng, 2009, p. 589), and the monetization of blogs is even less examined. By examining the objective relations that underlie the particular social positioning of bloggers, and those economic structures in operation which enable the process of monetization, we unpack the notion of the Lifestyle blog as an increasingly common, explanatory class in the wider practices of blogging in Malaysia. We understand the class not as something given, but as something constituted, or *to be done*, via networks and discursive activity.

Following boyd, we conceptualise blogging "as a diverse set of practices that result in the production of diverse content on top of a medium that we call blogs" (2006, para. 1). Personal bloggers tend to feel that their blog is an extension of their self, both as a cathartic means of self-exploration and a reflexive repository of their thoughts; "the conventional ethos [is] 'I blog for me'" (Reed, 2005, p. 237), which resonates with Giddens' discussion of self-actualization, whose "moral thread [...] is one of *authenticity* [...] based on 'being true to oneself'" (Giddens, 1991, p. 78; original emphasis).

Connected to self-actualization, successfully monetizing a Lifestyle blog is a realized strategy for the distribution of both economic and symbolic capital. So we ask the following: First, how do people, institutions, and technologies intersect to enable the monetization of blogs? Second, how does the criteria of authenticity for Malaysian bloggers relate to this possibility of monetizing personal online expression? And finally, what role does the boundary object of web analytics play, in this negotiation of both symbolic and economic capital?

In a first pass, we will apply the methodological agnosticism of actor-network theory (ANT) in tracing the emergence of a specific "blog advertising network" company[2]—known hereafter as "BlogAdNet."[3] We examine how it influences the interaction of other key actants in the relative stabilization of Malaysian blog networks, in particular through the "black-boxing" or objectification of authenticity. In a second pass, Bourdieuan field/capital analysis will be used to describe how the relative autonomy of the Malaysian field of blogging, constituted historically through the sociotechnical separation of online and offline fields of social interaction, is affected by heteronomous actors exchanging and recalibrating social, economic, and symbolic capital.

We conclude that web analytics are an important nexus for this activity, in the sense that on the one hand personal and Lifestyle bloggers self-understand their capacity for authentic expression through the metrics of unique visitors and their accumulating comments. But on the other hand, unique visitors can easily and subtly shift in register, to become understood by the blogger as a form of economic capital for the purposes of marketing and advertising. Ultimately, the Lifestyle blogger's position is differentially formed between these tensions, as they communicate their life experiences within a milieu of informational capitalism.

BLOGGING IN MALAYSIA

Blogging started in Malaysia around 2001, and by 2003 the formation of the online portal "Project Petaling Street"[4] marked a certain maturity in the field. At this point, the Malaysian field of blogging was more inward looking, resembling in part the autonomous field of scientists who "tend to have no possible clients other than their own competitors" (as cited in Eyal, 2006, p. 11). As the field grew, different genres reformed around themselves—the social-political (SoPo) bloggers, the personal bloggers, the tech bloggers, and others. The SoPo bloggers were the first to take on a more outward-focused outlook, as they explicitly set out to challenge government domination of the media. Political reactions and defamation cases initially brought blogs to mainstream attention, which was subsequently also paid to a variety of other blogging genres, particularly the personal and food blog (Tan & Zawawi, 2008, pp.

19–32). As the audiences for blogs expanded, bloggers became more aware of interest coming from non-bloggers.

A major shift in Malaysian blogging began in early 2007 with the founding of BlogAdNet. The founders understood that as isolated units, very few blogs had any potential for advertisers, but grouped together, they could be a fertile platform for Malaysian advertising. Two years later, BlogAdNet is running regular and successful campaigns for many national and international leading brands across around 100,000 registered blogs—mostly in Malaysia and Singapore, but also regionally. The monetization of blogs has become possible for all Malaysian bloggers: though only a tiny proportion (less than 0.01 percent) have been able to generate monthly incomes equivalent to those of senior managers,[5] many more have been able to accumulate small amounts of money and other rewards. BlogAdNet has provided many occasions for bloggers to meet up, and as in the commercialized medium of radio, will often negotiate promotional barters with companies for those in attendance (e.g., Vaccaro, 1997).

THE MALAYSIAN LIFESTYLE BLOG AS AN ACTOR-NETWORK

Turning to the descriptive methods of ANT is primarily useful for addressing the dynamics and concrete materialities of *technologies* within its sociological approach, encompassing both human and non-human actors (actants) who "*modify other actors* [actants] *through a series of trials that can be listed thanks to some experimental protocol*" (Latour, 2004, p. 75; original emphasis). The framework considers the entire complex assemblage of institutions, technologies, and practices that make up a phenomenon, in this case Lifestyle blogging. With its adoption we need to be wary of allowing technical analogies from the World Wide Web to creep into our definition of network. In the case of ANT, "[n]etwork is a concept…It is a tool to help describe something, not what is being described." (Latour, 2005, p. 131); an actor is a network, and vice versa (Abramson, 1998, p. 3). As Callon notes, "[i]t is a question of where the buck stops: Either you focus on the group itself…in which case you have an actor. Or you pass through it into the networks that lie beyond, and you have a simple intermediary" (1991, p. 142, in Abramson, 1998). A Malaysian Lifestyle blog is a hybrid—made up of a multitude of actants amongst which are the blog posts, hyperlink networks, bloggers, and commenters.

The personal blog derives from the diarist genre (McNeill, 2003), focusing on a person's life and thoughts. It was the most common genre reported in 2005 (Tan & Zawawi, 2008, p. 38), who also made an early mention of the Lifestyle blog genre. By early 2007 the Lifestyle genre had become firmly differentiated, as described by a leading Malaysian blogger:

I consider mine a Lifestyle blog, I tend to travel a lot, attend events, and I write about food every now and then, write about my opinion on certain things [...] what's cool, what's there to do, this sort of thing [...] So it's an evolvement of my previous blogging style which is personal blog. With personal blogs you rather write a lot more about what you think...about your family, your relationship, stuff like that I guess. (KappaBlogger, personal communication, August 5, 2008)[6]

For the purposes of this paper, a triadic cluster of main actants needs to be drawn out: the Lifestyle blog—and its epitomic product, the advertorial;[7] BlogAdNet, and the advertising clients. ANT is a "sociology of translation," and the interaction of these actants operates through a process of translation that corresponds to two levels. At the "first level...the translation of the social and natural sciences into a common register...[which] works to order and define the world...[and at the] second level...actors become the authors of other actors...and [work] to connect the networks of actors which hold the social world together." (Abramson, 1998, p. 3)

The predominant expectation of originality (52.2 percent), photos (50.1 percent), and the ability to comment (80.1 percent) shown in Table 1 suggests the first form of translation: blogging as a medium "translates" personal writing into a common sociotechnical register, forming the basis of associations amongst bloggers whose blog posts maintain the durable presence of the Malaysian blogosphere as an actor-network. Various automated processes (such as RSS feeds) facilitate this for the readers, and a contingent hierarchy develops based on the popularity of bloggers, as expressed in terms of readership and search engine ranking. This second translation is objectively obtained through automated "centres of calculation" (Latour, 1987, pp. 215–257). Readership is calculated through web analytics, and search engine ranking predominantly through the opaque algorithms of Google, based on hyperlink distribution and key words. To access web analytics, bloggers may use the "dashboard" interface provided by BlogAdNet, or one of many others available online. These not only provide raw readership data, but can also be used to gauge differences between posts by indicating search terms used to reach the blog post, referring hyperlinks, geolocation of readers, etc. Commercial actors also rely heavily on such data to segment and target audiences.

Table 1: Survey results—What should a blog have?[8]

n = 553	Strongly disagree	Somewhat disagree	Neutral	Somewhat agree	Strongly agree
"A blogger should only use their own original material"					
	4.7%	19.9%	23.1%	33.6%	18.6%
"A blog should have a comments function"					
	1.6%	4.3%	13.9%	37.6%	42.5%
"A blog should have photos"					
	4.7%	15.6%	29.7%	29.5%	20.6%

To successfully monetize blogs, BlogAdNet has to convince potential clients to place advertisements on them. The co-founder (also a blogger) remarks that one of the first hurdles is an ignorance of the medium:

> I give [the potential clients] two explanations: one's the technical one [a website with entries displayed in reverse chronological order]; and one is what it really is […] as where people share thoughts and opinions about any subject, or even on their own personal lives. (BetaBlogger, personal communication, February 2, 2009)

BlogAdNet also needs to translate collections of disparate blog posts into the terms of a conventional advertising register: demographics, size of readership, market segment, and so on. As BetaBlogger remarks, "you have to figure out a way to measure this hype to justify the clients' ROI [Return On Investment]" (op. cit.). Again, we can trace the means of translation to the web analytics: of particular relevance for the clients was the ability to pay only for those "Unique Visitors" (i.e., each reader is counted once every 24 hours only) from Malaysian Internet Protocol (IP) addresses. Other methods include getting registered bloggers to fill out surveys about their content and audience, and providing bloggers with technical means for collecting demographic details from readers. Using such tools and the web analytics, bloggers learn that by strategizing writing and linking practices, and getting more Malaysian readers, they can improve their earnings.

Based on these new metrics and strategies, the emergence of advertorials, being requisitioned and remunerated "thoughts and opinions" (BetaBlogger, op. cit.) (as opposed to simple banner advertisements), struck at the heart of the authentic blog. This was a very significant change, with their substantial earning power[9] introducing a strong power imbalance in favour of the clients and BlogAdNet. BlogAdNet has also successfully enrolled smaller bloggers, by providing a regular stream of free cinema tickets, free meals, parties, and contests—all centred on marketing events. It is clear that BlogAdNet has "become established as powerful through the stability of the networks that pass through them" (Couldry, 2004, p. 7). It would be easy at this point to portray the blogosphere as being "colonised" by BlogAdNet and their clients, but both ANT and field theory enjoin us to lay out all the jostling actants by viewing a flat panorama (Latour, 2005, p. 171).

Clients are drawn to the blogs because of the migration of younger affluent consumers away from traditional media forms. There have been attempts by companies to create blogs, but these have had mixed results (Kelleher, 2009), sometimes even backfiring on the company (e.g., Gogoi, 2006) because of perceptions of inauthentic manipulation. From this perspective, it is clear that the asymmetry engendered by the direction of the flow of income is reversed, and the Lifestyle blog has become a definitive "obligatory passage point" (Latour 1986, as cited in Abramson, 1998) for clients within or without BlogAdNet. Indeed, some clients have chosen to bypass them, engaging directly with the Lifestyle blogger.

Through these passage points in the network, actants "domesticate" other actants (Latour, 2004, p. 38). One such instance is described by GammaBlogger, in relation to the respective treatment of journalists and bloggers, wherein a cosmetics company organises:

> 'Intimate Sessions with Bloggers,' where they invite a few female bloggers to try their cosmetics [...] separating the bloggers and print media [...] they have to look at each blogger as a different publication, they're the editor of their own website [...] They have to be very individual [...] very personal, when they're dealing with bloggers; because everyone's personality is so different. (GammaBlogger, personal communication, May 7, 2009)

The regular events organized by BlogAdNet have an "actantial" (Abramson, 1998, p. 4), dimension as well, promoting a particular type of behaviour. They inscribe a narrative of community among formerly dispersed bloggers, and marry it to one of personal achievement—that is, come together, meet people you have encountered online, elicit reciprocal recognition, and record your co-presence through shared photos, enjoy common activities (movies, games, food, conversation) and win prizes—these are all embodied practices structuring the sociocultural order through the benefits of community. Prizes offer an incentive to blog about it afterwards, though many bloggers do in any case, believing it will make them more likely to be invited to the next event. Those Lifestyle bloggers who respond attract attention, readers and also imitators, and in this way a particular kind of narrative blog post, typical of the Lifestyle blog, is encouraged.

It is through these processes that the Lifestyle blog emerges as a new actant. Community activities and online diaristic practices fold into a holistic prototype for new bloggers entering the blogosphere, one which is suitably tailored as a mediator for the branding and advertising messages of clients. Some evidence for this is adduced in the survey, where bloggers who attend such events are more likely to self-categorise their blog as "Lifestyle"[10]; the most enterprising of them even using "Lifestyle Blogger" as a title on business cards. It demonstrates too that domestication can act in both directions. The first generation of Lifestyle bloggers, emerging out of a relatively autonomous field, have often been able to bend the clients to their needs. In a few cases, a blogger who was also a junior marketing or PR executive within a client's company took the initiative to introduce blogs to the company. But, as we will see, the autonomy of the field may be more difficult to maintain as newer Lifestyle bloggers arrive—emulating the practices and aspirations of current leading bloggers, but facing more limited economic success through lacking the leverage of loyal audiences.

The Genuine Article

These developments in Malaysian blogging have not been welcomed by all, with disquiet usually expressed in terms of blogs becoming too commercialised (Table 2—50.1 percent agreed with this), often focusing on whether the "disclosure" of advertorials is important (Table 2—63.7 percent preferred disclosure), that is, whether or not the blogger should explicitly state that they have been paid for the post, or given free goods as an incentive.

Table 2: Survey results—Opinions about blogs

n = 553	Strongly disagree	Somewhat disagree	Neutral	Somewhat agree	Strongly agree	Not applicable to me
"Blogs are too commercialised nowadays"						
	2.9%	15.9%	28.2%	31.3%	18.8%	2.9%
"Bloggers should be required to always clearly mark advertorials as such"						
	1.6%	2.5%	25.9%	31.3%	32.4%	6.3%
"Anyone can do what they like with their own blog"						
	4.2%	6.5%	8.0%	27.5%	46.5%	7.4%

When faced with disagreements, Latour's suggestion is to "deploy the full range of controversies about which associations are possible…[and] show through which means such controversies are settled and how such settlements are kept up" (Latour, 2005, p. 160). From the interviews conducted, the three main ways in which the issue is settled for bloggers are:

1. noting that disclosure is not legally required[11];
2. indicating an advertorial by using a tag;
3. stating the common assertion that "Everyone can do what they want with their blog" (Table 2—74 percent).

When probed, a fourth point will usually also appear: bloggers will assert that they will never say something they don't believe in, and that if they did, their readership would suffer. Similarly, a marketing executive for a distributor of a global sportswear brand said that they liked to use blogs because they can see the reaction of "real customers," from the bloggers and comments in the blogs. When asked if she was worried that by paying bloggers they may not get honest feedback, she also responded that if they failed to be genuinely sincere in their posts, they would lose their audience (personal communication, September 5, 2009).

A notable fourth actant that has not yet been discussed is "the audience," but due to the complexity of construing "the audience" as an actor-network, all of the facets of this collective of thousands of individuals obviously cannot be dealt with here. We can only (but still importantly) infer from the above remarks that it is constantly serving as a guarantor of authenticity for the blog.

This specific word, "authenticity," is most likely to be used in marketing discourse (e.g., van der Wolf, 2007), and academic debate (e.g., Brake, 2009; Lenhart, 2005). Recalling Giddens, for bloggers it encapsulates a kind of ethos, often coming in the form of advice to new bloggers to find their own style and "be yourself." This is reflected in the attitudes surveyed in Table 3: almost 75 percent believe that bloggers should always be honest; 62.2 percent expect to learn about the blogger as a person from their blog; and 69.4 percent would feel cheated if they discovered a blog was fictitious.

Table 3: Survey results—Blogs and authenticity

n = 553	Strongly disagree	Somewhat disagree	Neutral	Somewhat agree	Strongly agree
"When I read a blog, I should learn something about the blogger as a person"					
	2.4%	11.4%	24.1%	46.1%	16.1%
"A blogger should always be honest about his/her true feelings and thoughts"					
	1.8%	7.8%	15.6%	34.9%	40.0%
"If I found out that a blog that I had been reading regularly was completely made-up, I would feel cheated"					
	5.2%	9.8%	15.6%	28.4%	41.0%

Specific to the controversy about advertorials, the guarantee of authenticity is the backdrop against which a new association—a blog post with a directed commercial message—can become settled, or "stabilised" (Latour, 2005, p. 161). In an intimate rhetorical act of performative translation most prized by advertisers, the advertorial will be done in the trademark style of the blogger, perhaps also weaving it into a prosaic narrative of everyday life. Here, readers often assert that as long as the advertorials are done in the unique style of the blogger, they are satisfied. And the final guarantee is the "vote with your clicks" argument: 'If you don't want to read it, don't."

To get a sense of what's at stake in this fusion of diaristic authenticity with marketing messages, one can point to the notion of media "liveness" discussed by Couldry. He argues that "[t]he special status given to 'live' media can therefore be understood in actor-network terms as the time when media's status *as mediation* is most effectively black-boxed" (2004, p. 10; original emphasis). Just as the original success of CNN depended on the ability to transmit things "as they happen," bracketing editorial and practical gatekeeping processes, so too does the performative translation of authenticity disappear in the advertorial. In ANT terms it is "black-boxed," obscuring the moment where mediation is doing substantial ideological work, by designing strategies around blogging which mesh as closely as possible to the compositional experience of blogging and its particular communicative conditions of operation. To the extent that this is successful the message can be fitted in, preserving its authenticity.

But the difficulty in accounting for authenticity within an ANT framework is that it does not have a clearly stable, empirical referent. It leaves traces in practices, as a term cropping up occasionally in the idiom of blogging, and the answers to the survey do suggest that it operates as a category. But we argue next that field theory lends itself better to explaining how authenticity is best understood as a proxy term for *symbolic capital*, which is central to shaping the dispositions of Malaysian bloggers. We now examine this actor-network from the "inside," according to its social dynamics. We supplement the perceived deficiency of ANT with Bourdieuan field theory, opening up the black box to allow for a more nuanced description of the processes of discursive power involved.

MALAYSIAN BLOGGING AS BOURDIEUAN FIELD

ANT makes a set of theoretical commitments which tend to focus heavily on externalized relations between humans and non-humans. But to get some sense of the self-understanding of human actors within the assemblage, we can also turn to the field analysis of Pierre Bourdieu. With a reduced focus on materialities, key to his social-agential perspective are both the idea of *distinction*, and that of a market of *symbolic goods*. Under these alternative terms, we take the personal blog and the advertorial as examples of symbolic goods that circulate through networks of meaning online, hoping to impart a better theoretical sense to the aforementioned notion of being authentic. Illustrating the need to keep an advertorial unobtrusive, for example, AlphaBlogger explained to one commenter that, "It's not a good advert when it sounds like an advertorial"; and to another, justifying his placid response to some insults, "I would have ripped him another one if this post isn't what it is. :) [i.e. an advertorial] [...] I know this self-censorship isn't me, but hey, it pays the bills."[12]

Blogging is a kind of social sense-making. Through regularized entries, hyperlinking, and other classificatory schemes, bloggers express, organize, and perform their personally distinctive sense of the world. This personal distinction has a constant presence within social spaces, through the operation of *taste*: the generative ability to express and recognize a "distinctive lifestyle," that is, "the system of distinctive features which cannot fail to be perceived as a systematic expression of a particular class of conditions of existence" (Bourdieu 1984, p. 175).

We are concerned here with the "class" of bloggers, whose "practical knowledge of the relationships between distinctive signs and positions" (ibid.) enable them to form a framework for the composition of blog posts, while the blog itself accumulates an inscribed structure that represents a given person's taste over time. Personal blogs develop a distribution of positions along these lines in one way, and Lifestyle blogs another. That these posts are durably and publicly available online also represents the possibility of them being a source of potential *income* for the blogger, in a so-called "marketplace of attention" (Turow & Lokman, 2008, p. 23). As noted above, few bloggers are able to generate significant income, but the potential and the impetus provided by the example of the few that do incentivises and sharpens focus on the rationales for transitioning between a personal blog and a Lifestyle blog. A second set of concepts taken from Bourdieu is useful here.

Bourdieu divides the field of cultural production into two sectors: the field of *restricted* production and the field of *large-scale cultural* production, depending respectively on whether *symbolic* or *economic* considerations are foregrounded (Bourdieu 1985, p. 13). Following his analysis, it is helpful to understand blog entries here as analogous to public, artistic-literary *position-takings*, and not exclusively personal diaristic entries (Bourdieu 1993, p. 30). The field of restricted production has to do with cultural goods produced for consumption by a public of *producers*—in this case, other bloggers. In this field, adhering to the *illusio*—Bourdieu's term for the investment that compels each agent to take up a position within the cultural field—means investing in the idea that "I blog for me." This is paramount, as failing to do so ensures that the acknowledgment of a tasteful performance by other producers remains elusive. Against this restricted field, the *larger* field of cultural production is "destined for non-producers of cultural goods, 'the public at large'" (Bourdieu 1985, p. 17). So Malaysian bloggers understand their lifestyle in relation to other bloggers—most notably through structured blog comments at the end of each entry—and also to a general, mostly imagined, audience online.

Finally, we also need to get a brief sense of Bourdieu's notion of *symbolic capital*, a source of power which, "responds to socially constituted 'collective expectations' and beliefs" and influences "social agents endowed with the categories of perception and appreciation permitting them to perceive, know and recognize it" (Bourdieu 1998, p. 102). Its operation is limited by its recognition, "when the mental structures of the one to whom the injunction is addressed are in accordance with the structures inscribed in the injunction addressed to him." (ibid.) In other words, symbolic capital is how bloggers achieve *influence*, and it is through this influence that they might conceivably monetize their blog.

Symbolic capital is organized into goods through the medium of the blog. But as mentioned, these same goods can also circulate as *economic* capital, and here some of the dynamics further delimiting the difference between personal blogs and Lifestyle blogs emerge. Restricted cultural production mitigates position-taking and symbolic goods against monetization ('blogging for its and/or my own sake'), whereas large-scale cultural production requires "submission to external demand [...] characterized by the subordinate position of cultural producers in relation to the controllers of production and diffusion media [...]" (Bourdieu 1985, p. 28). ThetaBlogger, an openly gay blogger and employee of BlogAdNet, demonstrates these divergent concerns:

> I do give some thought to this [how people may see my blog], because technically I know that there are various audiences reading my blog [...] And, not to mention that my blog is actually *partially* commercialised. So it is important to make sure that it doesn't cross to the border, that I can't earn from ads, and at the same time I do not want to offend my readers who might be conservative [...] of course some things I do intend to share, to enlighten them what's the real world like, in the gay society, but I will not delve into the details, completely. (ThetaBlogger, personal communication, August 10, 2009; original emphasis)

In registering with BlogAdNet, and by taking the decision to include advertorials on their blogs, Lifestyle bloggers exchange symbolic autonomy within a field of restricted production (i.e., blogging) for a relative increase in economic autonomy within the larger-scale field of cultural production. This negotiated exchange of autonomy happens in light of thresholds determined through the position of the blogger, their writing style, and their web analytics. It happens through BlogAdNet by straddling the two fields of restricted and large-scale cultural production, and marking advertorials as "in, but not of" the Lifestyle blog, through carefully structured affiliations.

The exact *mode* of Lifestyle blogging as cultural production becomes a point of struggle, with bloggers adjusting the meaning of their entries by either accommodating them to positions in a field newly stabilized to include corporate and marketing clients, or defining themselves against such actors. The objective necessities of life are translated through taste into lifestyle strategies and structures of significance. Together the fields contribute to the ongoing reproduction of a *habitus*—"the durably installed generative principle of regulated improvisations" (Bourdieu 1977, p. 78)—for the Malaysian field of blogging. Visiting audiences eventually attuned to these new agents of exchange in the field adjust their expectations, while technological interfaces mediate the stabilizations in various ways. For example, blog reading software like Google Reader allows for the quick scanning of blogs, enabling the reader to ignore advertorials. In the space that remains for this section, through Bourdieu we will outline a more detailed conceptualization of these forces for a blogger, as they rationalize the monetization of their blog. On one side of the tension is the Lifestyle blog post as a symbolic good, and on the other is the Lifestyle blog post as an economic good.

Distinction: The Lifestyle Blog Post as a Symbolic Good

In terms of symbolic capital within an economy of blogging, there are always homologies between producers and consumers, which generate a field of possible position-takings. More simply put, people blog in the hope that their photos and writing will generate some kind of symbolic acknowledgement or response from readers—recognition. If such recognition is regularized, then the blogger accrues symbolic capital; comments publicly reinforce or challenge the blogger's taste, and visits observed through web analytics provide a symbolic index of their worth. The symbolic goods—blog entries, and the blog itself as a site—are thus unconsciously always adjusted and adjusting to the heteronomous expectations of the blogosphere overall, as a blogger achieves distinction and a larger audience from which to symbolically profit. This symbolic capital also comes to collectively yield *autonomy* for communal categories of bloggers through their cultural distinction, such as with the class of Lifestyle bloggers who have had success in distinguishing themselves from other styles of blogging, through BlogAdNet.

From this first view of the blog post, there is a *symbolic* hierarchization. For the blogger, a positive polarization of the symbolic is established via a negative distancing from economic profit; an indifference to that economy as it relates directly to blogging, and a positive striving for cultural recognition by one's peers and an audience. Distinction is achieved by negotiating the core terms of blogging itself noted earlier, its *illusio*.

Heteronomization: The Lifestyle Blog-Post as an Economic Good

Homologies also exist between producers (bloggers) and consumers (readers) vis-à-vis their position-takings towards externally dominant agents of economic power, and these positions bear on the blog posts. Bourdieu sees "lifestyle" as something very particular, and not some benign selection amongst the aisles of consumer fashion and technology—as a cursory look at a Lifestyle blog may suggest. It is rather an actively differentiated system of objective-material conditions for one's life, married to a coherent structure of significance for a given person (Bourdieu 1984, p. 172). Lifestyle actively differentiates us from others in the formation of identity. On the basis of these overarching conditions, we constantly misrecognize the objective-material truth of situations, and under Bourdieu's terms symbolically *euphemize* ourselves in the face of power. Consider the extract that follows, where we see how the blog has accompanied the blogger through her life. The blog's meaning changes as her lifestyle transforms from student life to working life. When money becomes a driving factor, the motivation to "keep it alive" shifts, with concern focusing more on the audience than on being "for myself."

> I've gone through my phases of writing; first it was, solely for myself; then I realised I had an audience, but it was OK, it's still for myself. And then [...] I was really busy in working life, but I was still trying to keep the blog alive [...] And, after that the advertising money started coming in, so it was important to keep it alive [...] I just felt like [...] I've got nothing to blog about! [...] Who wants to hear about me complaining about work? I don't! (ChiBlogger, personal communication, October 26, 2009.)

Euphemizing oneself means making an embodied-semiotic move that keeps the game of the social order going—one circulates symbolic, intellectual, or economic capital, which has the twin function of establishing one's subjectivity and lifestyle through a position-taking, while at the same time deflecting the objective reality of material asymmetries that constitute relations of power. The Lifestyle blog as a genre, and the advertorial as a style of post, may be best understood as *emerging from a latent attitude towards the conditions of euphemization that the medium of blogging produces*. Lifestyle bloggers euphemize the potential source of income by emergently formulating terms, conditions and polite ways of dealing with advertorials, so that they become rationalized into the social order, and can circulate amongst other bloggers (see Table 4). Thus, in Table 4, we see how the intention of bloggers is foregrounded—53.6 percent think that making money should not be the driving reason for blogging.

Table 4: Survey results—Blogs and monetization

n = 553	Strongly disagree	Somewhat disagree	Neutral	Somewhat agree	Strongly agree	Not applicable to me
"Making money from a blog may change it a bit, but not enough to make the change important."						
	3.1%	13.6%	32.9%	35.4%	11.0%	4.0%
"Advertisements don't change a blog, but advertorials do."						
	1.3%	8.5%	28.8%	36.5%	19.7%	5.2%
"If a blogger has a blog only in order to make money, it's not a real blog"						
	7.1%	17.9%	20.1%	25.0%	28.6%	1.4%

Understanding the Lifestyle blog post in this way, through an emphasis on economic goods, inverts the earlier terms of autonomization: a negative polarization is established through a *presumption* of accumulated symbolic recognition. Consequently a more positive polarity is set up on the part of the blogger towards a more functionalistic, "middle-brow" audience fitted to marketing needs. This is the central way that symbolic capital is converted into economic capital when it comes to blog entries; economic capital appropriates the *illusio* of blogging for the purposes of influence and viral marketing. Analogous to the idea of an "obligatory passage point" marked in the earlier ANT analysis, BlogAdNet serves as the passage point between the blog post as symbolic good, and the blog post as an economic good. It serves the dual purpose of both autonomization and heteronomization, by operating as a built authority for *consecrating* the activity of Lifestyle bloggers as a class.

BlogAdNet as a Consecrating Institution

First, BlogAdNet repositions the Lifestyle blog as a symbolic good in the field of advertising, where the blogger and their audience come to represent authentic consumers. It becomes a source of aggregated advertising space, with an umbrella statistical apparatus to back up its claims. Second, BlogAdNet helps to consecrate Lifestyle blogging as a specific *genre* of blogging. Third, it constructs an overarching economic system of relations, and a set of cultural dispositions between each of the categories of restricted producers (e.g., food blogging, or personal blogging), so that there is a new heteronomous consensus in support of blogging based in web analytics and keyword analysis, which deliver bloggers their potential income. BlogAdNet connects formerly disparate, restricted-production bloggers to the logic of an unrestricted public, through objectified demographic research, which as we have seen may have a subsequent impact on what each will put up on their respective blogs.

But at the same time it organizes Lifestyle blogging in this mercantile way, BlogAdNet also preserves the internal symbolic norms of blogging through a built sense of community, maintaining its cultural legitimacy even as it brackets and appropriates the larger economic motive. The opposition of power and money to the importance of authentic cultural expression, as elicited in the surveyed concerns regarding commercialization (Table 2), operates ultimately as a kind of co-existence in opposition (Bourdieu 1985, p. 31). There is no normative/foundational center here, and more the negotiation of a boundary; authenticity is dialogical in its preservation through exchange. Telling others about one's lifestyle (the internal function of Lifestyle blogging) always has an external, objective role in the wider economy, and BlogAdNet reflects this negotiation in operation. What is the objective basis upon which this boundary moves? The blog post, as both symbolic and economic good, is measurable through its objectification in web analytics.

FROM AUTONOMY TO HETERONOMY THROUGH THE NEXUS OF WEB ANALYTICS

Marking the importance of web analytics, and the ways they are interpreted by both bloggers and those involved in the attention economy, we reach an interesting nexus between actor-network theory and field theory. As an objective measure of both visits and the keywords used to reach one's blog, server logs deliver web analytics that act as a Latourian center of calculation: a metrical point of translation for bloggers to imagine their audience(s), and leverage monetization strategies. By quantifying a framework that can give at least tentative answers to these questions, web analytics enable the archetypal Malaysian blogger to shift back and forth between understanding their efforts as symbolic or economic. Web analytics thus serve as a "boundary object," objects which are "both adaptable to different viewpoints, and robust enough to maintain identity across them" (Star & Griesemer, 1989, p. 387). As Eyal argues, field theory is not well equipped to account for "boundary-work," but networks are able to "provide for a seamless connection between the two fields" (2006, p. 5). In this case, we can understand web analytics as the objective point of heteronomous articulation, "the obligatory point of passage," of the blog between the Personal genre in the autonomous field of blogging and the Lifestyle genre in the economic field.

Parenthetically, we might extend this point to Internet social science research itself. Increased computing power, visualisation software, and the ubiquity of hyperlinks have enabled powerful quantitative strategies for almost anyone who cares to access robust web analytics—lay person or expert—to objectively describe online networked relationality in a variety of ways. The reflexive researcher must account for the risk of substantializing statistically hyper-connected descriptions of this sort. Tallying up links, and letting representational strategies of quantification take over as a procedural proxy for what's "really going on" is an issue to be wary of. Both ANT and field theory offer complementary ways to deal with this issue. An ANT approach would maintain that these digital tools for parsing complexity are themselves mediators, and not simply intermediaries for what Adamic argues is "what had been hidden before, the social relationship (accessible to the social scientist only through time-consuming individual interviews)" (2008, p. 227); while Bourdieu's field theory usefully admonishes us to, at the same time, "stop thinking in terms of entities, proper names, concrete individuals, things, and begin grasping all of these as bundles of relations" (Eyal, 2006, p. 1). For Bourdieu, the latter are always hermeneutic, and genetic to a particular field of analysis. Relations of significance can become occluded through tools like web analytics, which tend to describe relations through an empirical-functional lens.

Conclusions

Considering that the audience is both the necessary cause of the advertisers' interest (needing to put their money where the eyes are) and the ultimate guarantor of the ineffable substance of authenticity, it is tempting to argue that they are in fact the final definite point of passage in these interconnected actor-networks. But that would be missing the point of ANT, that there is no obligatory centre anywhere (Latour, 2005, pp. 178–179): BlogAdNet, web analytics, authenticity, and the bloggers are all contingent centres.

Both Eyal (2006) and Couldry (2004) have argued that ANT does not properly account for power differentials; Latour argues in his defence that the "two tasks of *taking into account* and *putting into order* have to be kept separate." (2005, p. 207), and that ANT can only do the first task properly, setting the stage for an effective political engagement in the second. In the above analysis, we have tried to show that ANT reveals the substantive processes of translation and mediation that make up a Malaysian Lifestyle blog—a hybrid that grows more robust by enrolling other actants as it is reproduced, responding to their different needs through a particular mix of technology, personal expression, and economic incentive. Bourdieuan field theory filled out this account, giving a description of the processes of discursive power involved.

Through a strategic aggregation of web analytics, BlogAdNet has been shown to be a key mediator, translating the activities that constitute both the social and economic dimensions of the Lifestyle blog. It achieves this capacity for naturalizing the advertorial by translating the personal and fitting it into activities that can support advertising messages. Bourdieu has supplied the means by which to indicate the discursive forces at work in this naturalization by arguing that the trope of authenticity should be better understood as an ongoing function of particular exchanges of symbolic goods and capital. Expression fits into economic capital through immanent processes of distinction: taste-making on the part of the Lifestyle bloggers themselves. On this point we close reflexively, by arriving at another nexus.

Bloggers are agents who seek authenticity by using web analytics, objectifying it quantitatively for both symbolic and economic purposes. Doing so modifies their habitus, recursively defining their significance in reproduced discursive and material-economic networks. Faced with a welter of objective data now available from analytic tools designed to study the social order online, as they recursively create their object of study, can internet researchers say any less of themselves?

Notes

1. Participant-observation from 2007 to 2009, a survey conducted in March-April 2009, and extended interviews with bloggers (August 2008 to November 2009).
2. A 'blog advertising network' is a company that registers blogs and then brokers ad space on them.
3. All bloggers and the company have been given pseudonyms.
4. http://www.petalingstreet.org/ (last accessed February 15, 2010)

5. For international parallels see for example Thompson (2006)
6. The language used in the interviews is transcribed verbatim, without corrections into Standard English.
7. The advertorial developed out of spontaneous reviews of products, services, etc.
8. April 2009 online survey; 553 completed responses (356 bloggers and 197 readers). As it was a convenience sample, statistical significance was not calculated. More details at julianhopkins.net.
9. An advertorial can pay from RM200 to RM4500. A starting graduate salary is about RM2000.
10. 38.6 percent of those who attend such events listed 'Lifestyle' as one of their top three self-categorisations, compared to 20 percent of those who do not attend.
11. Recent Federal Trade Commission rules have rendered disclosure a legal requirement in the USA.
12. AlphaBlogger blog post, January 12, 2009. Details on request.

REFERENCES

Abramson, B. D. (1998). Translating nations: Actor-network theory in/and Canada. *The Canadian Review of Sociology and Anthropology, 35*(1), 1–19.

Adamic, L. A. (2008). The Social Hyperlink. In *The Hyperlinked Society: Questioning Connections in the Digital Age* (pp. 227–249). Ann Arbor: University of Michigan Press and University of Michigan Library. Retrieved January 9, 2010, from http://hdl.handle.net/2027/sp0.5680986.0001.001

Bourdieu, P. (1977). *Outline of a theory of practice*. Cambridge, UK: Cambridge University Press.

Bourdieu, P. (1984). *Distinction: A social critique of the judgement of taste*. Cambridge, MA: Harvard University Press.

Bourdieu, P. (1985). The market of symbolic goods. *Poetics, 14*(1), 13–44.

Bourdieu, P. (1993). *The field of cultural production. Essays on art and literature*. (R. Johnson, Ed.). Cambridge, UK: Polity.

Bourdieu, P. (1998). *Practical reason : On the theory of action*. Stanford, CA: Stanford University Press.

boyd, D. (2006). A blogger's blog: Exploring the definition of a medium. *Reconstruction: Studies in Contemporary Culture, 6*(4). Retrieved September 30, 2007, from http://reconstruction.eserver.org/064/boyd.shtml

Brake, D. R. (2009). 'As if nobody's reading?': The imagined audience and socio-technical biases in personal blogging practice in the UK. PhD, London School of Economics. Retrieved January 14, 2010, from http://eprints.lse.ac.uk/25535/

Cenite, M., Detenber, B. H., Koh, A. W., Lim, A. L. H., & Ng, E. S. (2009). Doing the right thing online: A survey of bloggers' ethical beliefs and practices. *New Media & Society, 11*(4), 575–597. doi: 10.1177/1461444809102961

Couldry, N. (2004). Actor network theory and media: Do they connect and on what terms? In A. E. A. Hepp (Ed.), *Cultures of Connectivity*. n/a: n/a. Retrieved January 17, 2007, from http://www.lse.ac.uk/collections/media@lse/pdf/Couldry/Couldry_ActorNetworkTheoryMedia.pdf

Eyal, G. (2006). Spaces between fields. Presented at the American sociological association, Montreal Convention Center, Montreal, Quebec, Canada. Retrieved September 6, 2009, from http://www.allacademic.com/meta/p94468_index.html

Giddens, A. (1991). *Modernity and self-identity. Self and society in the late modern age*. Stanford, CA: Stanford University Press.

Gogoi, P. (2006, October 9). Wal-Mart's Jim and Laura: The real story. *BusinessWeek: TopNews*. Retrieved October 21, 2009, from http://www.businessweek.com/print/bwdaily/dnflash/content/oct2006/db20061009_579137.htm

Kelleher, T. (2009). Conversational voice, communicated commitment, and public relations outcomes in interactive online communication. *Journal of Communication, 59*(1), 172–188.

Latour, B. (1987). *Science in action: How to follow scientists and engineers through society* (p. 274). Cambridge, MA: Harvard University Press.

Latour, B. (2004). *Politics of nature: How to bring the sciences into democracy*. (C. Porter, Tran.). Cambridge, MA: Harvard University Press.

Latour, B. (2005). *Reassembling the social: An introduction to actor-network-theory*. Oxford: Oxford University Press.

Lenhart, A. B. (2005, April 21). *Unstable texts: An ethnographic look at how bloggers and their audience negotiate self-presentation, authenticity, and norm formation*. Master of Arts in Communication, Culture and Technology , Georgetown University. Retrieved March 25, 2008, from http://cct.georgetown.edu/7904.html

McNeill, L. (2003). Teaching an old genre new tricks: The diary on the internet.*Biography, 26*(1), 24–47.

Reed, A. (2005). 'My Blog Is Me': Texts and Persons in UK Online Journal Culture (and Anthropology). *Ethnos, 70*(2), 220–242. doi: http://dx.doi.org/10.1080/00141840500141311

Star, S. L., & Griesemer, J. R. (1989). Institutional Ecology, 'Translations' and Boundary Objects: Amateurs and Professionals in Berkeley's Museum of Vertebrate Zoology, 1907-39. *Social Studies of Science, 19*(3), 387–420. doi: 10.1177/030631289019003001

Tan, J., & Zawawi, I. (2008). *Blogging and democratization in Malaysia: A new civil society in the making*. Kuala Lumpur, Malaysia: SIRD.

Thompson, C. (2006, February 12). Blogs to riches—The haves and have-nots of the blogging boom. *New York Magazine*. Retrieved February 10, 2010, from http://nymag.com/news/media/15967/

Turow, J., & Lokman, T. (2008). *The hyperlinked society: Questioning connections in the digital age*. Ann Arbor: University of Michigan Press and University of Michigan Library. Retrieved January 9, 2010, from http://hdl.handle.net/2027/sp0.5680986.0001.001

Vaccaro, J. P. (1997). The prevalence of barter in radio. *Journal of Promotion Management, 4*(1), 27–37.

van der Wolf, M. (2007). *The business value of blogging*. Retrieved December 29, 2007, from http://www.lewispr.com/Business_value_of_blogging.pdf

CHAPTER EIGHT

Cyberinfrastructure Inside Out

Definition and Influences Shaping Its Emergence, Development, and Implementation in the Early 21st Century

KERK KEE, LUCY CRADDUCK, BRIDGET BLODGETT & RAMI OLWAN

INTRODUCTION

Cyberinfrastructure (Atkins, et al., 2003; Seidel, Muñoz, Meacham, & Whitson, 2009; Stewart, 2007), also commonly known as e-infrastructure in the United Kingdom (Meyer & Dutton, 2009 ; Meyer, Schroeder, & Dutton, 2008) and e-research infrastructure in Australia and Europe (Eccles et al., 2009; Jankowski, 2009; Schroeder, 2007a), officially emerged and was recognized at the turn of the millennium. Since then, it has attracted serious attention and much investment from the scientific and scholarly communities as an emerging method and platform for research; and from political and policy institutions as a new entity with tremendous economic, societal, and global implications. Due to its potential, multiple stakeholder groups are grappling with the concept of cyberinfrastructure and engaging in the building of this "next-generation Internet" (Foster, Kesselman, & Tuecke, 2001, p. 217). As we look forward to the second decade of the 21st century, the time is ripe to explore three interrelated research questions:

- ~ What is cyberinfrastructure?
- ~ What are they key political influences shaping its domestic emergence and development?
- ~ What are the key challenges impacting its international implementation?

By drawing from literature in the social sciences, law, and policy studies, as well as computer and information sciences, this chapter attempts to provide some preliminary answers to these important questions.

Our purpose for this chapter is threefold. First, for scholars from a variety of disciplines and policymakers to study cyberinfrastructure, it is critical to have a coherent definition. Therefore, we synthesize definitions from a range of disciplines and propose an integrated and generative definition of cyberinfrastructure based on four dimensions: characteristics, layers, processes, and outcomes. As cyberinfrastructure continues to expand in the future, we anticipate that these four dimensions will generate new examples while remaining useful as an integrated framework.

Second, for appropriate policies to be developed around cyberinfrastructure, we sketch a political model of cyberinfrastructure development based on three key influences: market, policy, and law. By drawing upon the case of the Internet, this model describes the recursive relationships among these three domestic influences and how they can impact the case of cyberinfrastructure. We also discuss the concepts of digital divides and network neutrality as cautionary tales from the Internet that scholars and policymakers need to keep in mind while they work to advance cyberinfrastructure domestically.

Third, as cyberinfrastructure builds on the Internet, its implementation at the international scale deserves careful examination. Cyberinfrastructure projects at national borders often encounter challenges that limit their effective implementation. We sketch a typology based on key influences in three categories: international challenges, national challenges, and common project challenges. We argue that the lack of an international standard policy structure is a fundamental challenge to effective implementation. We then discuss how these influences shape cyberinfrastructure implementation at the international scale while balancing between policy harmonization and conformation among large and small countries.

Collectively, we believe the integrated definition, the political model of domestic influences, and the typology of international influences contribute to an understanding of the emergence, development, and implementation of cyberinfrastructure in the early 21st century. Although the chapter is primarily based on developments in, and the jurisdictions of, the United States and Australia, some reference is made to positions and/or laws of the European Union and the United Kingdom to ensure a more comprehensive discussion. We conclude the chapter with a brief discussion of implications and a modest proposal for future research. Let us turn our attention to the first research question: What is cyberinfrastructure?

DEFINING CYBERINFRASTRUCTURE

Although scholars have reviewed the definitions of related concepts, such as *e-science* (Jankowski, 2007; Schroeder, 2007b) and *collaboratory* (Olson, Zimmerman, &

Bos, 2008), the definition of *cyberinfrastructure* is still unclear. Scholars in computer and information science, science and technology studies, communication, sociology, and management have written about cyberinfrastructure based on a diverse range of disciplinary perspectives and research agendas. Instead of picking the most cited definition, we believe a synthesis of existing definitions serves to bring several important insights together. We propose an integrated and generative definition of cyberinfrastructure based on its characteristics, layers, processes, and outcomes. Table 1 provides a preview of the discussion to follow.

CHARACTERISTICS

Table 1: Elements of the generative definition of cyberinfrastructure

Dimensions	Descriptions	Elements
Characteristics	Inherent properties of cyberinfrastructure	Data-intensive Computationally powerful Distributed, hierarchical Interoperable Second-order growth
Layers	Important components for the creation of cyberinfrastructure	Hardware Software Agents Interactions
Processes	Key functional operations of cyberinfrastructure	Virtual environment Virtual organization
Outcomes	Results of cyberinfrastructure implementation	Productivity Innovation Revolution

Cyberinfrastructure can be characterized as data-intensive, computationally powerful, distributed, hierarchical, interoperable, and with second-order growth (i.e., generation of data about data and metadata). The first characteristic of data-intensiveness refers to cyberinfrastructure's capacity to hold a large amount of data in various forms, including numbers, text, multimedia, acoustic, and nonverbal data (Poole, 2009). The goal of combining data sets among groups of researchers was a key driver of initial cyberinfrastructure development. Traditionally, science was limited by regional data, human resources, and the technological capacity of small groups of independent researchers at various locations. With cyberinfrastructure, researchers can combine multiple datasets into one that exceeds what a traditional small group of researchers can collect and analyze. Consequently, researchers can undertake research at a scale otherwise not possible.

Cyberinfrastructure is computationally powerful and has the capacity to analyze intensive data (Friendlander, 2008) via parallel and distributed computing processes. Traditionally, researchers executed computer analyses of scientific data in local laboratories, and research studies were limited by the processing speed and power of individual (or a small network of) commercial personal computers (PCs) or local supercomputers, if available at affiliated institutions. A supercomputer is a large network of powerful modular servers and commercial PCs run on parallel and distributed computing algorithms. Via this technique, a data-intensive job can be divided into small chunks and fed into individual servers and/or PCs (within a supercomputer architecture) concurrently and recursively. The results are aggregated at the end of the computational process, thus increasing processing speed and capacity. Cyberinfrastructure is a network of supercomputers across the country, such as TeraGrid, which connects 11 supercomputers across the United States. Due to the combined computational power, cyberinfrastructure provides the fastest computational resources available, enabling science at a speed otherwise not possible.

As alluded to in the characteristic of computational power, cyberinfrastructure is a distributed platform. Via cyberinfrastructure, a group of researchers can submit a computationally and data-intensive job from a remote location, and have the job processed at multiple supercomputers, with the combined results returned back to the initiating location. The group of researchers is only required to have access to cyberinfrastructure through their local institution. The distributed characteristic of cyberinfrastructure takes research beyond local constraints to virtual computational resources at impressive speed.

Due to its complexity, it is logical to describe cyberinfrastructure as hierarchical. Cyberinfrastructure involves a range of large and small components from a cable modem that can be picked up by a child to a supercomputer the size of a building basement. However, it is important to note that since cyberinfrastructure is a network it cannot function properly without its smallest component (Friendlander, 2008). Therefore, the smallest component also holds the entire infrastructure together, although it may be hierarchical in a physical sense.

In order for cyberinfrastructure to operate and function as a coherent whole, the scientific data, computational resources, technological systems, and human organizations must be interoperable. Interoperability (ACLS, 2009; Baker, Ribes, Millerand, & Bowker, 2005) refers to a property of cyberinfrastructure wherein a range of diverse data, resources, systems, and organizations interoperate and work seamlessly together. Without interoperability, the former four characteristics reviewed have no value. Data sets, computer resources, computational jobs, and technological components will remain local, separate, and small-scale. Given the aforementioned characteristics, cyberinfrastructure data and its subsequent analysis grow over time, leading to second-order growth. The aggregated scientific data and the human activities recorded on cyberinfrastructure are first-order data that can be analyzed by researchers. Careful coding, qualitative observations, statistical analyses, and network visualizations by these researchers yield second-order data (Poole, 2009) or metadata added to the existing cyberinfrastructure data repositories. This unique characteristic facilitates longitudinal research in a wide range of disciplines at a scale and fashion never possible before. Cyberinfrastructure grows in its potential for new discoveries by means of complex cross-referencing (Poole, 2009) to explore large-scale global challenges.

In sum, cyberinfrastructure possesses the characteristics of data-intensiveness, computational power, distribution, hierarchy, interoperability, and with second-order growth. Its complex make-up requires a careful explication of its different layers.

Layers

The second dimension of cyberinfrastructure is constituted by its four layers of hardware, software, agents, and interactions. The hardware layer can be further divided into the specialized/niche hardware and the general/commercial layers. Based on the discussion thus far, it is apparent that a key piece of the cyberinfrastructure puzzle is a network of supercomputers. Supercomputers are mainly used for niche research analyses as described earlier, and specialized commercial applications, such as airplane design, automotive crash tests, and oil reservoir discovery, too big in scale and expensive for frequent trials and errors.

The specialized/niche layer of cyberinfrastructure also includes advanced instruments (Stewart, 2007), digitally-enabled sensors, observatories and experimental facilities (NSF, 2007), and large-scale data storage systems/repositories (Atkins et al., 2003). These are examples of a range of physical hardware for specialized purposes and niche usage. Many of them, such as observatories, were uniquely built with specific utilizations in mind. In other words, these instruments are hardware that cannot simply be bought "off-the-shelf."

Beyond its specialized/niche hardware, cyberinfrastructure is also made up of a general/commercial layer of distributed personal computers (Atkins et al., 2003), desktops (Friendlander, 2008), and portals (Poole, 2009). In addition to commercial PCs, cyberinfrastructure also includes phone devices (landline, mobile, and smart phones), fax machines, printers, modems, and other off-the-shelf electronic devices researchers concurrently use for nonresearch purposes. These hardware components, both specialized/niche and general/commercial, are tied together through a range of software applications.

The software layer mirrors the hardware layer in terms of its specialized and general applications. In order to process large-scale data and specialized analyses on supercomputers, researchers need appropriate analytic tools (Poole, 2009) and high-performance computing (HPC) applications. HPC applications are used by highly trained researchers. Loosely generalized, HPC applications enable parallel and distributed processing of large-scale data on supercomputers to generate scientific results. However, these results need to be shared with collaborating researchers and interested colleagues. This is where the next category of software applications comes in.

A range of information and communication technologies (ICTs) supported by telecommunication systems, the Internet, and the World Wide Web make up another critical layer of cyberinfrastructure. Specific examples include email applications, online meetings, personal and organizational web pages, digital libraries, search engines such as Google (Hai, 2004), and web 2.0 technologies such as blogs (Poole, 2009). These ICTs can be used by researchers for interpersonal, group, and organizational communication between their scientific and nonscientific work concurrently.

We suggest that individual or collective human agents are key to an active cyberinfrastructure. Without users, something that is called "a cyberinfrastructure" is not a real cyberinfrastructure (Ribes & Finholt, 2009), as it is not active but static. The notions of people (Stewart, 2007), groups and organizations (Lee, Dourish, & Mark, 2006), and personnel and institutions (Atkins et al., 2003) have consistently been mentioned in cyberinfrastructure literature. Human agents usually are assumed to be independent actors in the context of cyberinfrastructure. That is, they represent "nodes" in a network, as understood in traditional social network literature.

Next to human agents, nonhuman agents are also important in cyberinfrastructure. Nonhuman agents refer to documents, concepts, key words, data sets, etc. (Contractor, 2009). They are discrete entities and resources (Friendlander, 2008) in the context of cyberinfrastructure. However, they are labeled as "agents" because they appear to do things, or have impacts on other "nodes" in the network. The notion of nonhuman agents departs from traditional social network literature and draws upon actor network theory (Latour, 2005). However, human and nonhuman agents as nodes have no impact on each other or the overall cyberinfrastructure if they do not interact.

Human and nonhuman agents interact and are tied together through multidimensional networks. The notion of networks concerns not only high-performance grid networks (Stewart, 2007) and the Internet in the physical sense, but relationships and ties as commonly defined in social network literature. Furthermore, the notion of networks is "multidimensional," because nodes in cyberinfrastructure are both people and "nonhuman agents" (Contractor, 2009). Networks therefore represent complex physical connections and relational ties among human and nonhuman agents in cyberinfrastructure.

Middleware is a specific type of multidimensional network (NSF, 2007). It is composed of computer software that ties multiple software applications together and allows them to interact in a parallel, distributed, and interoperable environment. We highlight middleware because it plays a significant role in creating key processes in cyberinfrastructure, which will be discussed next. Cyberinfrastructure consists of hardware, software, agents, and interactions. When the four layers are in actions, they create cyberinfrastructure processes.

Processes

There are two key processes of cyberinfrastructure: the virtual environment and the virtual organization. The first key cyberinfrastructure process is the technologically generated virtual environment (VE) (ACLS, 2009; Poole, 2009; Schroeder & Axelsson, 2006), which represents the continuously generated virtual space in which researchers interact with data and each other. In the present development, virtual environment consists of visualizations (Stewart, 2007), simulations (Leonardi, 2009), and models (Monteiro & Keating, 2009). Based on HPC applications on large-scale data, researchers are able to use visualization techniques, interactive simulations, and computer modeling to analyze and predict complex scientific phenomena with significant societal and global implications. One example is real-time simulation on combined data from nearby locations that effect the development and expansion of a hurricane threatening a local community.

The second key cyberinfrastructure process is the socially generated virtual organization (VO). A VO brings a group of distributed researchers together for a common purpose and allows them to interact with each other. A VO is dispersed, diverse, and flexible while also remaining coordinated, coherent, and secured (Bird, Jones, & Kee, 2009). For instance, the VO for the Large Hadron Collider project brings about 2,000 researchers together across multiple countries. Embedded with the notion of a VO is interdisciplinary collaboration (Monteiro & Keating, 2009) and community-building (Poole, 2009). So far, cyberinfrastructure can be defined by its unique characteristics, multiple layers, and key processes; however, it is most importantly defined by its intended outcomes.

Outcomes

Cyberinfrastructure emerged to promote three specific outcomes. The first outcome of cyberinfrastructure is that it increases productivity (Stewart, 2007). Productivity can be understood as the ability to do more in less time. Due to its intensive data and computational power, cyberinfrastructure can process larger data at faster speed than traditional personal computers and local networked machines. If nothing else, cyberinfrastructure increases the productivity of researchers.

The second outcome of cyberinfrastructure is innovation (Atkins et al., 2003). Innovation refers to the ability to produce novel outcomes. Due to its intensive data and computational power, cyberinfrastructure enables research at a scale and speed never before possible, facilitating the exploration of complex phenomena at the edge of scientific frontiers. As a result, cyberinfrastructure enables and supports innovations in research.

The third outcome of cyberinfrastructure is that of revolution (Atkins et al., 2003; Stewart, 2007). Revolution can be defined as the ability to cause a paradigm shift. With increased productivity and a stream of innovations, cyberinfrastructure stimulates a revolution in science, causing researchers to think of science differently, explore big questions, and work in new ways. Cyberinfrastructure also generates a new set of scientific practices (Monteiro & Keating, 2009). Once transitioned, researchers cannot return to their previous paradigm of doing science, thus effecting a intellectual and practical revolution.

INTEGRATED AND GENERATIVE DEFINITION OF CYBERINFRASTRUCTURE

Taken together, cyberinfrastructure is data-intensive, computationally powerful, large-scale, distributed, hierarchical, interoperable, and with second-order growth over time. It consists of specialized and general hardware, high-performance computing applications and information and communication technologies, human and nonhuman agents, all interacting and connecting through multidimensional networks. This platform facilitates technologically generated virtual environments and socially generated virtual organizations that orient people, data, and technology towards common goals. Cyberinfrastructure leads to increased productivity, breakthrough innovations, and paradigmatic revolutions. Simplistically, cyberinfrastructure is an empowering network of advanced technologies, metadata, and collaborative individuals and groups.

At the heart of cyberinfrastructure's outcomes of productivity, innovation, and revolution is the notion of "empowerment." Empowerment can be interpreted as the ability to mobilize people to produce, innovate, and revolutionize. In the influential Blue Ribbons Report, which led the US National Science Foundation to eventually establish the Office of Cyberinfrastructure in 2005, Atkins and colleagues (2003) articulate that cyberinfrastructure "should provide an effective and efficient platform for the empowerment of specific communities of researchers to innovate and eventually revolutionize what they do, how they do it, and who participates" (p.5). This vision is ambitious and bold, and reminds us of the initial emergence of the Internet as an information and communication network for scientists to share information and collaborate.

Following its public introduction, the Internet has become a political and economic phenomenon where multiple stakeholders wrestle to define what people do, how they do it, and who participates. Beyond proposing an integrated and generative definition of cyberinfrastructure, we intend to discuss the domestic and international influences that shape cyberinfrastructure's emergence in the early 21st century. The second research question we explore is: what are the key political influences shaping its domestic emergence and development?

Proposing a Model of Domestic Influences

The emergence, development, and implementation of cyberinfrastructure are primarily a national effort. In this section, we propose a political model of domestic influences on cyberinfrastructure based on three interrelated forces: market, policy, and law, including the differing impact of "soft" laws (i.e., accepted policies or industry behaviors) and "hard" laws (i.e., as found in legislation) as shown in Figure 1, p. 175). We pay particular attention to the issues of digital divides (Hammond, 2002; Ypsilanti & Paltridge, 2004) and network neutrality (Wu, 2003) in the context of the Internet, and access to it and its services, because they have significant implications for the future market, policy, and law surrounding cyberinfrastructure. Whilst cyberinfrastructure effectively has no comparator, lessons may be learnt from the development of the Internet that will benefit the future development of cyberinfrastructure.

Digital technologies have radically changed how people interact, work, learn, engage politically, and spend their free time. As Benkler (2006) considers, "[t]he change brought about by the networked information environment is deep" (p. 1). Whilst the development of computers from being a research tool to their wide adoption for private use occurred over several decades (Bresnan & Greenstein, 1999), it is anticipated that cyberinfrastructure adoption for commercial and private use will occur more quickly.

Two Critical Issues of the Current Internet

Digital Divides

The term "digital divide" has been used by authors to refer to a variety of gaps. As Gunkel (2003) considers, these gaps include those representing the inequality in educational opportunities; the differences of opinions regarding engineering solutions; and the level of access to new technologies and a person's ability to use them. The digital divides of relevance to this chapter are those comprised of the gaps between people who have effective access to and are able to utilize the Internet and those who do not. These gaps are often defined through the categorization of users by their socioeconomic status, generation, and education levels. Although digital divides exist between countries at the global level, we primarily focus on their domestic manifestation.

The Internet now enables a wealth of information to be available to all at the touch of a key. One divide that exists is that not all people in all countries are able, for reasons of government policy or their own lack of means; or in some cases lack of desire (Crump & McIlroy, 2003), to access this information. Given the ubiquity of the Internet and information technologies, there is the risk that these people will be disadvantaged in the future without the access to digital information for the education, work, and social activities and engagement (Crump & McIlroy, 2003).

The ability to access information by itself is but one divide. Another of relevance for this chapter concerns the distinction between those who are digitally literate and those who are not. However, having the ability to access information is not the same as being literate, which to an extent involves the ability to communicate with others (Cole & Lorch, 2003) beyond the abilities to read and write (Buckingham, 2007). This is an important distinction to draw for digital literacy.

To be digitally literate, similarly to being literate generally, is more than simply being able to access information on the Internet. Utilizing Prensky's (2001) terminology for the sake of ease of description, whilst appreciating that it is a starting point and merely representative of broader and more complex issues, we suggest that "digital immigrants" must quickly become digitally literate if they are to do more than merely exist in the 21st century. They need to acquire the communication norms and social rules of cyberspace in order to coexist with the "digital natives" (Prensky, 2001) for whom digital information and computer-mediated communication are an integral part of everyday life. In other words, digital immigrants must learn how to study digitally; to read and contribute to blogs effectively; to send tweets appropriately; to visit virtual worlds competently; and to create and read texts and emails meaningfully.

Prior claims that education will exponentially and significantly improve with the introduction of the Internet (Jankowski, 2007) have not in fact been realized. They will not be realized until everyone is able to access it from anywhere with digital literacy (Tapia, Blodgett, & Jang, 2009). Beyond education, the digital divide also restricts access to vital governmental services, leading to what some have referred to as the "participation divide" (Goldfinch, Gauld, & Herbison, 2009 p. 335). Therefore, the need to boost participation, as identified by La Rose et al (2007), rises to the top of the policy agenda in a 21st century society. How is the lesson of the digital divide applicable to the case of cyberinfrastructure?

The key implication of the current digital divide for cyberinfrastructure is how to avoid creating another form of (digital and participation) divide while at the same time ensuring the current one is overcome without sacrificing the needs of all users in the current system (Tapia et al., 2009). In the case of cyberinfrastructure, literacy involves acquiring the skills and knowledge to remotely access supercomputers and high-performance computing applications with advanced computer programming techniques and computational thinking, along with the skills to manipulate large-scale data. Given the specialized layer of cyberinfrastructure defined earlier, most digital immigrants and digital natives of the Internet are considered "immigrants" to cyberinfrastructure. As the specialized layer of cyberinfrastructure continues to develop rapidly, a second-degree digital divide is likely to emerge. That is, the ability and/or means to use cyberinfrastructure, or rather the lack thereof, will impact not just ordinary users but also sophisticated ones without the necessary means or skills. A second-degree digital divide will have a detrimental impact on the vision of cyberinfrastructure as an effective and efficient platform for the empowerment of what people do, how they do it, and who participates, beginning in and with the scientific community. Given this anticipation, appropriate access and educational policies need to be developed to prepare and speed immigration of all users to cyberinfrastructure in the 21st century.

Within the United States, an example of such policy building has emerged in the form of the Broadband Technologies Opportunities Program drafted as part of the economic stimulus bill initiated by President Obama (Tapia et al., 2009). However, as Tapia et al. establish, there are many competing groups for limited cyberinfrastructure funds, which may require the creation of a twofold government policy. In addition to distributing funds to research and scientific groups for cyberinfrastructure projects, consideration must also be given to increasing the ability of all individuals to be connected to high-speed broadband Internet. An important aspect of overcoming the digital divide and ensuring widespread and ongoing access to a network is to ensure that government policy appropriately addresses the issue of network neutrality (Endres, 2009).

NETWORK NEUTRALITY

The issue of network neutrality gained serious public attention in the United States in 2002, about the same time as cyberinfrastructure started to gain momentum in the scientific and research community. However, network neutrality has been debated for several years in the United States, the United Kingdom, Germany, Italy, and Japan with little to no connection with the parallel cyberinfrastructure development.

As the issue has progressed, many policymakers and lawyers are interested in finding appropriate solutions for network neutrality in the context of the Internet.

Although its complexity makes definition difficult (Cave & Crocioni, 2007), network neutrality can be defined as preventing Internet providers from blocking, speeding, or slowing web content based on its source, ownership, or destination (savetheinternet.com, 2009). Currently, network operators around the world engage in various discriminatory behaviors to control what is being sent over the Internet. New technologies are able to discriminate between different applications; that is, Internet carriers will examine the packets (data) sent, see what applications it comes from and tier users' access based on that analysis. For example, one way this discrimination may occur is that a "black box" may be installed in the network as a packet sniffing technology in order to recognize and decode certain packets of interest within network traffic (Dierickx, 2006) for the purposes of selective deprioritization. A more simplistic description is that the network operator performs "content filtering based on the type of data being sent" (Kariyawasam, 2007, p. 101). A lack of network neutrality therefore means service providers (who are also access providers) can act to prevent users' access to their competitors' services; or can work to make that access less effective; or can "…create a charging structure based around protocol usage" (Kariyawasam, 2007, p. 102).

At the heart of the debate is the "openness" of the Internet for those seeking lawful access to contents and services. Network neutrality simply requires network operators not to distinguish between data packets, whether in the form of a text, video, chat or any other format, and to push them through their pipelines at the same speed (Editorial, 2009). Network operators such as telecommunication and cable companies argue that they should be able to provide preferential treatment to online companies willing to pay for their data packages to be transferred faster than others. The profit from such arrangements will allow telecommunication and cable companies to further develop advanced fiber-optic networks and increase broadband access to more users. Moreover, they argue that discrimination is needed to protect their users against spam and other security threats, and to insure the quality of VOIP services (Cerf, 2006). These arguments constitute the principle of network diversity.

However, there are three arguments for network neutrality and against network diversity in the form of discriminatory control by network operators. First, network neutrality prevents anticompetitive practices by cable and telecommunications companies, which enjoy a domination of the market (Berner-Lee, 2006). Second, network neutrality will help promote Internet innovation, no matter how big or small, by allowing everyone to be a creator, speaker, and broadcaster (Balkin, 2006). Third, enforcing neutrality ensures the free flow of information and will prevent the evolution of a two-tiered system in which service providers could inhibit or prioritize the transmission of data based on what is good for their own business, not what is in the users' best interests (Editorial, 2009). The network should treat all content, sites, and platforms equally (Wu, 2003). The lack of network neutrality presents a serious and real threat to the Internet and its future "as an open network" (Berner-Lee, 2006).

The open architecture of the Internet allows users to publish their work without payment of fees, and without seeking permission from anyone. This openness however is changing with network diversity that differentiates between packets and changes the underlying architectural design of the Internet. Just as the Internet rests on foundations of openness, most technology is not developed in isolation as any new technology builds on prior technologies/applications. The suggestion therefore is that a completely independent technology is not possible (Nakamura, 2000) as without the openness of the Internet for sharing, learning, and experimenting, many of the services we currently take for granted would soon cease to exist (Johnson, 2009) and new services may never develop in the first place.

Instead of debating the principles of network neutrality or network diversity, Zittrain (2006) argues for looking at these principles as a "means" in and of themselves rather than an "end." He challenges common understandings of the architectural design of the Internet and focuses on fundamental issues associated with upholding principles of openness. Lessig and McChesney identify some of these key issues:

> Most of the great innovators in the history of the Internet started out in their garages with great ideas and little capital. This is no accident. Network neutrality protections minimized control by the network owners, maximized competition and invited outsiders in to innovate. Net neutrality guaranteed a free and competitive market for Internet content. The benefits are extraordinary and undeniable (...). (Lessig & McChesney, 2006)

This argument for a free and competitive market that fosters innovation can be extended to cyberinfrastructure. As defined earlier, cyberinfrastructure ideally envisions an effective and efficient platform for the empowerment of what people do, how they do it, and who participates. Currently, cyberinfrastructure operations, such as in the case of TeraGrid in the United States, are supported by multiple academic supercomputer centers and national research laboratories. In the case of the Enabling Grids for E-sciencE (EGEE) in Europe, it is supported by many supercomputer centers across multiple countries. Many of these supercomputer centers are supported by governmental funding and cyberinfrastructure is not yet commercialized. However, commercial and corporate involvements in cyberinfrastructure development and implementation are increasing. Drawing from the parallel comparison between the Internet and cyberinfrastructure, appropriate funding and commercial policies need to be developed to ensure an open cyberinfrastructure platform with minimum or no domination or monopoly by specific supercomputer centers, access operators, or countries.

The choices we make today in connection with how we should run the Internet network are critical for its future and that of cyberinfrastructure. Appropriate access, educational, funding, and commercial policies need to be developed to reduce and possibly avoid the issues of digital divide and network neutrality being replicated with the case of cyberinfrastructure. In order to help guide this preventative effort, we propose a political model of how market, policy, and law may regulate the emergence and development of cyberinfrastructure.

A Political Model of Domestic Market, Policy, and Law

Our examination of market, policy, and law and how they can impact the case of cyberinfrastructure begins with a consideration of three fundamental issues that must be kept in mind. First, in most societies, disputes and conflicts of interests are resolved in terms of norms and standards. In making any decision, reliance is placed by the parties involved on the rules and principles provided by statutes and precedents (Boulle, 1996). As stated earlier, although cyberinfrastructure has no comparator, lessons from the development of the Internet deserve consideration. The issues of the Internet digital divides and network neutrality, and how they are dealt with by the market, policy, and law, will directly influence how cyberinfrastructure will be regulated by these. However, how courts interpret and apply the law (which tends to be focused on their specific jurisdiction only) will also influence cyberinfrastructure policy development, as judicial interpretations, and thus precedents will vary from jurisdiction to jurisdiction.

Second, the preferred method for policy and legislation development is to be proactive in dealing with an issue before it develops, such as the projected implications of the digital divide and network neutrality on cyberinfrastructure. However, governments more often than not take a reactive approach to the development of policy and legislation, doing so in a manner that Beardsley and Farrell (2005, p. 2) refer to as "trial and error [by] confusing economic goals with political and social ones." A reactive approach can, regretfully, allow an issue to continue well after it was first identified. Also, without a "global legislator" (Benvenisti, 2008), it is left to the separate jurisdictions to determine what appropriate cyberinfrastructure policy is and to develop and implement it through domestic law. This is a concern, as there is a real risk that there will be inconsistent policy adoption and implementation which may adversely impact upon cyberinfrastructure's future developments and deployments. How various governments perceive their position in developing and creating laws, and determining which entities are to be regulated by those laws, is a real concern, as there is a potential for conflict (Burk, 2007). In particular, the potential issues of the digital divide and network neutrality for cyberinfrastructure should be clearly identified and proactively addressed by early policy development to enable them to be resolved before they become issues in fact.

Third, changes in society are continual (Gibbs, 2000) and these changes affect both the courts (Cranston, 1986) and the law, in that policy, and thereafter the law, generally develops in response to those changes (Gibbs, 2000). One issue that requires specific consideration for any cyberinfrastructure project is in relation to the ownership, and the right to use any resource created. One solution may be to determine that the resources created would be made available to all by means of open licensing (Fitzgerald & Pappalardo, 2008). However, this is unlikely to be a workable option in practice, due to the interests of funding bodies (David & Spence, 2008). Other legal issues arise with respect to the legal relationships between the parties; the need for the apportionment of liability for any risks arising from the project: and how this is dealt with by the various domestic laws, as well as the need to ensure compliance with antitrust laws (David & Spence, 2003, 2008).

A concern is to ensure the ability of policy and laws to easily change and develop as society does (White, 2008). For cyberinfrastructure, this requires ensuring that any law and policy is both internationally consistent and also technology neutral, so as to encompass all future developments and thus maximize the prospects of the law's enforcement. This would be difficult to achieve practically but, even if it was achieved, another challenge arises. As the society within each jurisdiction is unique, any changes wrought by one society, which then require implementation into policy and law, are likely to vary from jurisdiction to jurisdiction. This could lead to a situation where whilst an activity is currently treated consistently by all jurisdictions, in the future, as has occurred with some intellectual property laws (Middleton, 2008), the same activity could be permitted in one jurisdiction but prohibited in another.

As Unsworth (2008) identifies, cyberinfrastructure "...is the infrastructure for a knowledge economy..." (p. 40) but it is also both "...a scientific challenge [and]...a social and human challenge ..." (p. 42). Therefore, in order to create appropriate and durable policy and law for cyberinfrastructure, it is necessary to ensure that policy is not created in isolation. That is, it must *not* be created by one jurisdiction only, in isolation from true input from all other jurisdictions; *nor* created by one interest group only in isolation from true input from all stakeholders. Proper cyberinfrastructure policy and law development will require a multi-disciplinary, multi-jurisdictional, open, accountable, and cooperative process. Policy development is also influenced by, and in turn influences, the operation of the market and the law. We propose a model detailing this relationship as follows: market drives policy, policy drives law, and law drives the market.

Market Drives Policy

The demand for appropriate infrastructure and policy generally is driven by demand for the services, although there are exceptions to this observation (such as in the case of Google or Facebook, where the service itself creates its own market; and in the case when specialist advice is required after taxation policy and laws have been changed, therefore the policy itself creates the need for the service). When there is a market demand for services, restrictions on consumer choice and thus on network neutrality can arise both by means of technological prevention and/or by means of contractual obligation. Policymakers and regulators are therefore required to balance the benefits technology brings against possible misappropriations to decide how and if choices should be restricted in order to address what others refer to as "…'High Tech' competition technology" (Depypere, 1995).

Moreover, the market is crucial in imposing a simultaneous constraint upon how an individual might behave in cyberspace through the price they exact (Lessig, 2009). An example of this market mechanism is the price of software constraining the ability of "netizens" to use it on the Internet and communicate with others. Lessig (1999) has identified fours modalities of regulating behaviors in cyberspace: through law, norms, market, and code (architecture), or any combination of them.

On the other hand, public policy is based on consistent principles and supported by enduring values in the society (CEDA, 2006), but is often influenced by market demands at the same time. In order to incorporate these principles and values, policy development should involve all relevant parties and follow a clear policy framework, which is implemented in the market rigorously and systematically by means of the adoption of clear and accountable processes (Edwards, 2001, p. 3). Edwards suggests a useful framework adapted from the Bridgman and Davis model. This adapted policy development framework utilizes the following six stages: issues identification (problem definition and articulation), policy analysis (data/information collection, objectives/questions clarification, and options/proposals development), consultation, decision, implementation, and evaluation.

Under the influence of the market, the process of policy and legislative implementation can be costly and time consuming (CEDA, 2006) as it may be necessary to revisit earlier policy development stages in Edward's (2001) framework before it is possible to move forward. In addition to the direct influence of the market, the "…electoral cycle can play a large part in determining what items get on the agenda and whether they are pursued past a certain point" (Edwards, 2001, p. 10). A clear example of this, and the impact that a change of government has on policy, was seen in Australia after the 2007 federal election, when significant aspects of the previous coalition government's broadband policy and projects (for example the OPEL contract) were "discontinued" (Department of Broadband, 2008).

Overall, therefore, the economic market, along with cyclical political influence, drives policy and therefore although, currently, there is a limited market for cyberinfrastructure, the economic market will be an influence in the future. We argue that the same interrelationship between market and policy will apply in the future in respect of cyberinfrastructure development and attempts by legislatures to regulate for it.

Policy Drives Law

The ultimate object of policy is the creation of a norm by which societal behavior is regulated, that is, the creation of law. As Holland (2006) explains, the objects of "…[l]aw are the creation and protection of legal rights," which he defines to be the "…capacity residing in one man of controlling with the assent and assistance of the State, the actions of others." (p. 66). Law making is a process that involves a variety of actors, including "…government ministers, and public servants, as well as experts such as academics and others in the community" (Edwards, 2001, p. 1). Law is essentially the ultimate implementation of developed policy into practice.

Furthermore, written law (i.e., that created by the legislature or parliament as opposed to judge-made law) is not developed in isolation. It requires the impetus of government or society and usually arises from the need to address something that is "missing" (i.e., not addressed by current law) or an issue has arisen since existing laws were written, and not addressed by those laws (Heydon's Case, 1584, p. 637). The law does not exist in a vacuum, and as such, any proper analysis requires that a law and its underlying policies be examined where, and why, it operates (Murray, 2007). A good starting point is to ask—what is the purpose of this law? The policy process therefore commences with an accurate identification of the objectives to be addressed (Edwards, 2001, p. 2). In this regard, policy drives law and will do so for cyberinfrastructure.

Law Drives the Market

In return, the law affects the market as a modality of regulating cyberspace, by using taxes to increase or reduce market constraints on certain behaviors and activities. The market could play an important role in regulating cyberinfrastructure when and where laws are not comprehensively put in place to regulate it. Initially this may not appear to be such a problem, as most cyberinfrastructure is funded by government; however, learning from the experience of the development of the Internet, after cyberinfrastructure becomes a public domain (even though this may be many years off) attempts at regulation are likely to be highly ineffectual. Furthermore, governments are also concerned with the enforceability of the laws they implement (Burmeister, 1999; Coughlan, Currie, Kindred, & Scassa, 2006), as without an effective means of enforcement, any law implemented is arguably not worth the paper it is printed on, is of no use to regulators, and of no comfort to consumers.

However, the influence of the law on the cyberinfrastructure market is, for the time being, different from the impact of the law on the Internet market. That is because cyberinfrastructure is a bespoke (mostly for e-science) and artisan (sometimes for digital humanities) product due to its newness, whereas the Internet is a commodity product. The Internet became widely diffused because personal computers and home computing suddenly became much cheaper and therefore more available to both businesses and consumers and networking capability was achieved (Crandall & Jackson, 2001). Moreover, there are many service operators and suppliers for Internet access at very cost effective prices. This ease of availability impacts on network neutrality, as consumers expect and demand a level of access unrelated to the ISP used. If consumers want to change the commodity used, they are able to easily change service operators and suppliers.

Conversely, cyberinfrastructure is so new and specialized that it is not yet possible to "buy it off the shelf." Cyberinfrastructure is a bespoke/artisan product that is yet to be commoditized and commercialized, and have its services mass-produced and as such has not yet reached the networking capability of the Internet. How we approach policy development and legal regulation for cyberinfrastructure needs to be undertaken differently and, it is suggested, with more forethought and wider consultation. Nonetheless, when everyone is able to easily access cyberinfrastructure enabled processes such as virtual environments and virtual organizations through a portal, as defined as a part of cyberinfrastructure's general layer, we are likely to witness the commoditization of cyberinfrastructure portals and services in a similar fashion to that of the Internet.

For many jurisdictions, digital divides and network neutrality are critical issues that shed light on the rapid development and future implementation of cyberinfrastructure as we move into the second decade of the 21st century. Appropriate and proactive policies, because they are influenced by market and the law, need to be developed with forethought and wide consultation to reduce or prevent similar impacts and ensure openness and empowerment. In the meantime, we must also consider that any overregulation could restrict (or prevent) creativity.

As depicted in Figure 1, issues relevant to network neutrality may be disruptive during the process of law making (as considered earlier) and thus influence the "shape" of the market. Similarly, but this time in respect of market use/adoption, issues of the digital divides, and whether or not these are appropriately addressed, will be disruptive by impacting on what policies are preferred and thus adopted. Finally, changes in political control, as considered earlier, may be disruptive by preventing long standing policies from being brought to fruition.

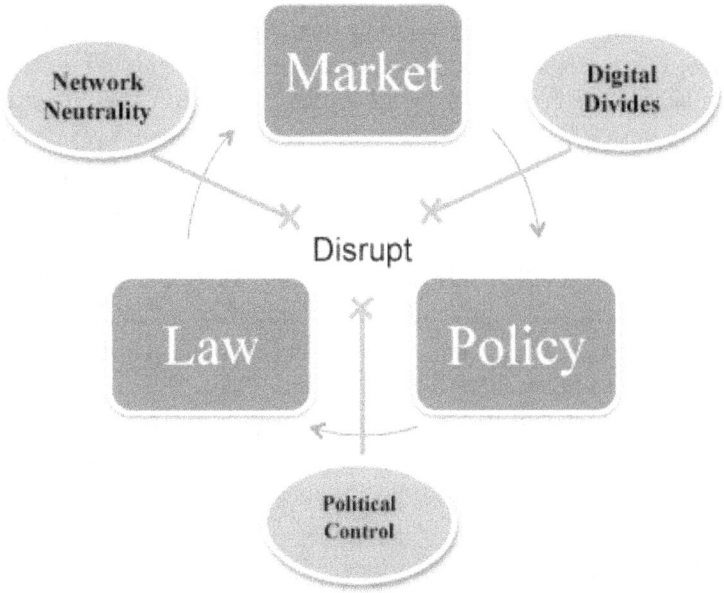

Figure 1. A political model of operation with moderating cyberinfrastructure factors

We have described a political model of cyberinfrastructure implementation based on market, policy, and law. It is important to note that where the conduct to be regulated or the product to be protected will have an impact on the international stage, it also requires international collective policy making. Moreover, from an international perspective herein lies a problem, as although the behavior constituting a breach of law may not fundamentally change from jurisdiction to jurisdiction, each jurisdiction likely identifies or describes the same act in a slightly different manner, and consequently legislates in a slightly different fashion to address it. These international aspects highlight the need to expand the discussion to the international arena. As such, what are the key challenges impacting cyberinfrastructure's international implementation?

Describing a Typology of International Influences

Cyberinfrastructure is not an isolated national effort. Given its rapid emergence and steady development, the vision of an all-encompassing digital system evolving into an "E-topia" (Mitchell, 2000) and "global innovation system" (Schroeder, 2007b, p. 3) may soon be witnessed internationally, as evidenced by the fact that the grid network infrastructure in Europe spans more than 30 countries (Bird et al., 2009).

Further, there is much investment internationally in cyberinfrastructure. The US National Science Foundation established the Office of Cyberinfrastructure in 2005 (Seidel et al., 2009) and the US government allocates about $175 million annually to develop and maintain its national cyberinfrastructure (Edwards, Jackson, Bowker, & Williams, 2009). The UK's Office of Science and Technology implemented a large funding initiative in 2000, and spent about £275 million between 2001 and 2006 in a similar effort (Edwards et al., 2009). The Australian e-infrastructure investment plan, represented by the Platforms for Collaboration capability area, had a notional $75 million allocated to it by the end of 2006, out of the total NCRIS budget of $542 million (Reid, 2007). The vision driving these investments is cyberinfrastructure's potential to improve cutting-edge research and enable global collaborations (Fry & Schroeder, 2009).

However, whilst cyberinfrastructure development clearly has government support, as identified previously, there is not currently an international policy, or an international regulator, or an international legislature addressing its development. Cyberinfrastructure policy and legal development is therefore left to the individual jurisdictions to manage by themselves, or preferably by means of international treaties and/or cooperation. The power of individual governments to create policy and laws, and the processes that they must follow, will impact upon what is ultimately (able to be) developed.

For example, the Australian government has the power to make laws for peace, order, and good government within their jurisdiction (Constitution, Sec. 51), and its power to legislate, as Gleeson CJ observes, "…includes a power to make laws with respect to places, persons, matter or things…external to—Australia" (XYV v Cth, 2005, p. 499). Proposed legislation and subordinate regulations are required to be approved by both the House of Representatives and Australian Senate before they become a law. The power in the United States, however, is found in both the legislative Congress and the executive branch headed by the President, which provide many of the same facilities as the Australian government. However, these initiatives are often undefined until approved by both the congress and president. Before this occurs they remain open in their scope, allowing for many of the details regarding the implementation of the law to be worked out by the state and local governments, or through government-run organizational branches such as the National Science

Foundation or National Telecommunications and Information Administration (Tapia et al., 2009). This could lead to uncertainty and inconsistent policy development.

Aside from the issue of the lack of one consistent policy and law, the issue of how an international cyberinfrastructure project is to be properly regulated is of concern where actions and actors are located in multiple jurisdictions (Burk, 2007), and any alleged breach of law may occur in one or many of those jurisdictions. For example, an Australian court may restrain conduct occurring outside the territorial boundaries of its jurisdiction; however, whether it does so is a matter of discretion for the court (Helicopter Utilities, 1963; Dunlop Rubber, 1921; Tosier, 1885). In exercising this discretion, Australian courts must determine whether they are a "clearly appropriate" forum (Voth, 1990) in which to determine the matter. Further, even if the court determines it is the appropriate forum within which to determine the matter, it may be that it will not grant an order that the court knows is unenforceable (Macquarie v Berg, 1999). Conversely, in the United States, the court must consider "…whether Congress intended extraterritorial application of the statute proscribing the alleged conduct" (Messigian, 2006). Due to the nature of operation of the American federal government, any extraterritorial actions require not only approval from the presidential seat but also from the congressional legislature. Even then, such actions and decisions are subject to review by the federal court, to determine if the actions taken are both within the realm of established common law as well as supportable under the Constitution, to ensure that any actions meet not only the founding ideals of the nation but also promote and support cooperative habitation with other states.

Although it may be difficult to achieve (as identified earlier), harmonization of policy and laws regarding cyberinfrastructure may be one means to address any international issues that arise from inconsistent policy development or legal application. The EU, for example, embraces harmonization of laws through Article 5 of the EC Treaty where, by means of the principle of subsidiarity, the EC is able to adopt measures at a community level where the objective of a regulation cannot be sufficiently achieved by member States, and where, by reason of the transnational nature of the offence, regulation can be better achieved at the community level. However, as can be seen from the European experience, working towards harmonization of policy and laws can take some time, and is not always successful in practice.[1]

However, the EU appears to have achieved a level of harmonization with regards to cyberinfrastructure. In June 2009, the European Commission, pursuant to Article 171 of the EC Treaty which gives the community power to "…set up joint undertakings or any other structure necessary for the efficient execution of Community research, technological development and demonstration programmes," adopted a specific regulation to enable the establishment of a European research infrastructure consortia (referred to as an ERIC) ("Community Legal Framework for a European Research Infrastructure Consortium (ERIC)," 2009). The establishment of this legal framework gives us confidence that, with regards to cyberinfrastructure at least, many of the international challenges identified earlier can and will be overcome in the not too distant future.

However, Murray's (2007) observation regarding the impact of any one state's laws on harmonization generally remains pertinent, in that "...a distinctive set of legal principles in any one nation can undermine the effectiveness of law as a regulatory tool in an international environment...." In order for there to be true harmonization of policies internationally and not just within the EU, all countries must have the same approach to their regulation of the actors and their multi-jurisdictional activities.

The desire for consistent policy and application must also be considered in the light of a recent observation by Benvenisti (2008) that some governments are increasingly "...consciously try[ing] to disengage from traditional law," in that they prefer informal means of commitment as opposed to establishing formal international organizations and treaties. Finally, others negatively (but perhaps realistically) view internationalization as an "...Americanization of the law..." (Michaels & Jansen, 2007). We are not taking a position on this matter but raise it simply as an observation from the literature. Essentially, we present a tension between harmonization and conformity at the international policy arena without advocating for a preference. It is simply an inherent tension that needs to be managed appropriately.

Additionally, despite much enthusiasm, cyberinfrastructure projects face international challenges, which are compounded in implementation by national specific issues, and common project challenges which are compounded in practice by project specific issues. We present a typology of challenges based on these three general categories with eight specific examples. A breakdown of each of these challenges can be found in Table 2.

Table 2: A typology of cyberinfrastructure challenges at several levels

International challenges	National challenges	Common project challenges
Lack of standard implementation policy	National funding for international projects	Lack of scientific communication practices
Country-based data storage location for an international project	National security concerns	Incompatible practices for internal knowledge management
Technological diversity across participating countries		Multidisciplinary backgrounds of the researchers

INTERNATIONAL CHALLENGES

The first international challenge involves working with technological diversity and choices that must be determined before cyberinfrastructure projects go forward. International cyberinfrastructure projects can be adversely impacted by differences in the speed and data carrying capacity of national communication backbones, which vary greatly from one nation to another—even in countries that share a border (Olson et al., 2008; Petrazzini & Kibati, 1999). Software used also often varies greatly between one country and another, as well as the hardware that is available to run current or developed cyberinfrastructure (Ackerman, Hofer, & Hanisch, 2008; Cyberinfrastructure Research Taskforce, 2005). International projects can be further complicated by the requirements set by national governments regarding the types of software that must be used in their research labs and governmental agencies. For example, Venezuela has an orientation towards open source software (Maldonado & Tapia, 2007).

The second challenge for international projects arises from the difficulties surrounding data storage for data produced or collected during an international scientific project (Arzberger et al., 2004). Currently, data in such projects is often stored on the site that has the faster connection or largest data storage capacity (Hofer, McKee, Birnholtz, & Avery, 2008). This arrangement creates a bias in favor of more developed nations when collaborations span very diverse countries, as it allows more developed nations to dictate how the data is stored, the times and methods through which it may be accessed, and generally privileges those countries in terms of travel and accessibility benefits.

The third challenge to international implementation is the lack of a standard implementation policy. There are many cyberinfrastructure projects spanning national borders that need to work with the continuing struggle of setting policies about how the project is coordinated, how credit and work are distributed, how risk is managed, how results are owned/shared, and how international and national regulations are handled. Often, each project is responsible for answering these questions for itself. Currently there are no set international policies regarding collaboration on a multinational project that engages with cyberinfrastructure (Lynch, 2008). However, while these policies do not currently exist, there are international initiatives, many arising from collaborations like the Large Hadron Collider (LHC), regarding the creation of such a policy. Two things that many of the participants in these primordial international policies seek to address are the issues of the digital divides and net neutrality that were raised earlier in this chapter (Lynch, 2008).

NATIONAL CHALLENGES

The fourth overall and first national specific challenge is imposed by national security concerns and the resulting increased costs to address the various concerns. For example, cyberinfrastructure project data often requires particular devices and software that are not accessible in areas outside of a single nation. Transportation of agents from one location to another can incur additional costs to an international project. In many cases, negotiations between the different agencies and governments involved in the project must be opened to discuss issues of visas and national security issues (Arzberger et al., 2004). In some cases, workarounds to this problem have been found through the further implementation or careful expansion of the current cyberinfrastructure, to allow for greater remote access to equipment and data (Ackerman et al., 2008; Myer, 2008; Olson et al., 2008). However, this workaround arrangement is not yet standard in international cyberinfrastructure projects.

The fifth challenge involves the funding arrangement adopted. International projects are often collaborations which draw their funding from various political entities such as the United Nations, countries like the United States or France, and multinational governments such as the European Union (Borgman, Wallis, Mayernik, & Pepe, 2007). Each of these entities wish to gain something from the results of the research, as well as position itself well in the international community as centers of science and research (Borgman et al., 2007; Hofer et al., 2008). This can put further pressures on international cyberinfrastructure projects as they attempt to include the entities' funding requirements in the goals and basis of the research project (Cyberinfrastructure Research Taskforce, 2005). This could result in any number of changes and can cause some project decisions (such as what software to use, how and where to store data, where research centers are located, and how time at the centers is divided), to become political decisions of national and international significance.

Although these five challenges involving technological choices, data storage, national security, and funding arrangement are particularly salient in national cyberinfrastructure projects where projects are influenced by the participants' desires, similar issues may also arise at the international level, as these projects are influenced by the participants' home jurisdictions' desires. In other words, such dynamics manifest themselves in most national and international cyberinfrastructure projects, although they appear magnified at the international level. The effects of these challenges are further compounded by the three common project challenges in our third category.

Common Project Challenges

The sixth challenge overall and first common project challenge is related to scientific communication practices, such as the setting of standards for communication, interaction, and documentation in international cyberinfrastructure projects. Scientific research groups often improvise standards for routine communication. However, these standards must be negotiated between the different members and power structures that exist in distant teams (Lee & Tibbo, 2007). A common problem cited by such project members is the passing of data and papers from one team member to another. Currently, teams have a wide variety of technologies to choose from, such as concurrent versions system (CVS), email, files, and file servers, to name just a few. Who is allowed to access and work on data, or publish from particular subsets of data must also be discussed to avoid complications (Lee & Tibbo, 2007). This can often be another downfall of data within projects being centrally located at one team's site, establishing them as gatekeepers who can determine what projects or information should be handed out or published, and which members of the team should be allowed to do so (Fry & Schroeder, 2009).

The seventh challenge involves internal knowledge management. The nature of many international cyberinfrastructure projects also elevates the need for compatible documentation policies and technologies, that is, interoperability of systems and processes. Since these projects can often span for many years, much longer than a single graduate student, research assistant, or even investigator may stay at an organization, documentation makes it possible for the knowledge, process, and policies established during the course of the project to be transferred from one researcher to another as the working staff of the project changes. However, international differences can often make the keeping of such documentation difficult (Cyberinfrastructure Research Taskforce, 2005). Preferences for the type of document technology used (text files, data repositories, videos, lab notes, etc.), as well as cultural idiosyncrasies (beyond the scope of this chapter to consider), can often render this valuable information difficult to access or understand (due to variances in dates, times, measurements, what information is recorded, etc.) (Lynch, 2008).

The eighth challenge arises in respect to the background of the researchers involved. Many CI projects are composed of researchers from different areas and disciplines of research (Lynch, 2008). This can further complicate existing tensions within international teams, as groups within the project vie to establish their interests as dominant among the research goals. A classic example given by Myer (2008) is the competition between computer scientists wishing to study the development of the CI programs used in the Pacific Northwest National Laboratory and the physicists who were using the CI to study energy. The tension that arises in such groups often exists because each group is invested in the project for a specific set of goals. In Myer's example, the computer scientists were most invested in pushing the boundaries of advanced remote presence and collaboration technology and software, while the physicists wished for an established and reliable set of tools which they could then use to focus on their own goals of studying energy consumption and usage.

Each project faces a number of unique difficulties that arise out of the combination of national contexts, institutions, researchers, and goals that make up that particular project. Table 2 summarizes some of the most common challenges faced by international cyberinfrastructure projects.

While many of these issues must be addressed on a one-by-one basis, the establishment of international standards and policies for collaboration on large-scale scientific research and the creation of cyberinfrastructure that specifically addresses these issues would ease the process of establishing a project for many future efforts.

Conclusion and Implications

As set out in the beginning of this chapter, the time is ripe to explore three interrelated research questions: What is cyberinfrastructure; what are they key policy influences shaping its domestic emergence and development; and what are the key challenges impacting its international implementation? We provided some preliminary answers to these important questions.

What Is Cyberinfrastructure?

Cyberinfrastructure is data-intensive, computationally powerful, large-scale, distributed, hierarchical, interoperable, and with second-order growth over time. It consists of specialized and general hardware, high-performance computing applications and information and communication technologies, and human and nonhuman agents, all of which interact and are connected through multi-dimensional networks. This platform facilitates technologically generated virtual environments and socially generated virtual organizations that orient people, data, and technology towards common goals. Cyberinfrastructure leads to increased productivity, breakthrough innovations, and paradigmatic revolutions. In short, cyberinfrastructure is an empowering network of advanced technologies, metadata, and collaborative people and groups.

What Are the Key Political Influences Shaping Cyberinfrastructure's Domestic Emergence and Development?

The political process through which domestic policies are made is complex. We propose a cyclical model to describe how the market drives policy; how policy drives law; and how law drives the market in return. This model helps us understand the key domestic influences shaping cyberinfrastructure emergence and development, especially with regard to the issues of the digital divides and network neutrality.

In the case of cyberinfrastructure, the digital divides present the tension between the wish to advance technology on the edge without creating another layer of access and participation division. Network neutrality reveals a further tension between the need to build more advanced and secure infrastructure without limiting creativity and innovation. Both issues have critical implications for cyberinfrastructure's vision of empowering what people do, how they do it, and who participates. Based on observations and lessons learned from the case of the Internet, we suggest early and proactive policies in the area of access, education, funding, and commercialization, to reduce and possibly avoid the effects of the digital divides and network neutrality in the case of cyberinfrastructure.

What Are the Key Barriers Impacting Its International Implementation?

We describe a typology of international challenges based on three categories: international specific challenges, nationally based challenges, and common project challenges. Key challenges include technological diversity and choices; international data storage decisions; lack of standard implementation policy; national security concerns; and funding arrangements. These challenges are amplified by three common project challenges: inconsistent scientific communication practices, incompatible internal knowledge management strategies, and diverging disciplinary interests. These challenges collectively point to the fundamental challenge of a lack of international standard policy to guide cyberinfrastructure projects at the global scale. In practice, there is no "global legislator" in the world. International cyberinfrastructure projects will have to balance the tensions of harmonization and conformation with the domestic law of big and small countries.

FUTURE RESEARCH

As we look into the future of cyberinfrastructure, there are nine key focal points we believe would advance our understanding of this important development. First, research could document cyberinfrastructure emergence, investigating which social, economical, political, and technological forces collectively led to the emergence of cyberinfrastructure in the early 21st century. Second, research could explore the design of cyberinfrastructure, especially the co-production between scientists as users and computational technologists as developers, as they co-design different pieces of cyberinfrastructure together. Third and similar to design-focused research, we could also pursue the process of development, especially with regards to the organization of cyberinfrastructure projects and related socio-technical dimensions of development, as they involve stakeholders such as funding agencies, supercomputer centers, policy institutions, and commercial vendors, in addition to users and developers.

Fourth, the process of adoption at both the individual and organizational levels deserves critical attention, as an infrastructure without individual and organizational users would be a major problem. Fifth, and in the beginning stage of adoption, research could track how cyberinfrastructure is used to support implementation at the micro level, such as distributed collaborations among teams of scientists and users. As the number of users grows and distributed collaborations begin to overlap, research could address the sixth focal point of virtual organizing and/or virtual organizations at the macro level of deployment.

Seventh, future research could also track the impacts of cyberinfrastructure adoption and implementation on individuals, groups, organizations, communities, societies, and the world. In the next few years, as the concept of cyberinfrastructure continues to emerge, the eighth focal point suggests exploration of the roles of supercomputer centers as service providers, infrastructure operators, and access regulators in open science, similar to the roles of Internet Service Providers (ISPs) in the case of the Internet and the information world. Ninth, and finally, when access to cyberinfrastructure becomes possible through commercial portals and cyberinfrastructure's funding mechanisms go beyond primarily governmental investments, future research could investigate the commercialization of cyberinfrastructure services as a public commodity beyond being a bespoke/artisan product.

Predicting the future of cyberinfrastructure and related research is problematic, as is most IT predictions, particularly in view of the specific and common challenges we have identified. Cyberinfrastructure is a complex and constantly developing phenomenon. Future researchers need to take into consideration the historical, political, and cultural context in which it has and is developing. What is clear, however, is that the future of cyberinfrastructure will be more certain if nations and disciplines work together to achieve it, as this multi-national, multi-disciplinary, and multi-time zone team have worked together to write this chapter.

Notes

- With thanks and acknowledgments to Ralph Schroeder for taking time over his Christmas break to provide comments on a prior version; Tessa Jade Houghton, our section Editor, for her invaluable, honest, and supportive critique; Yana Breindl, our secondary Editor, for her holistic comments and oversight; Andrea Tapia for her review and suggestions for improvements. All errors remain those of the authors.
- The authors are listed purely in the order in which they joined the chapter project.
1. For the progress of the European Parliament towards the creation of one area of freedom, "security and justice" see—European Parliament "Scoreboard: Union-wide fight against crime," committee on citizens; freedoms and rights, justice and home affairs, freedom, security, and justice: AN AGENDA FOR EUROPE, <http://www.europarl.europa.eu/comparl/libe/elsj/scoreboard/crime/default_en.htm> (last accessed January 10, 2010)

Cases and Legislation Cited

Commonwealth of Australia Constitution Act 1900

Dunlop Rubber Company v Dunlop [1921] 1 AC 367

Helicopter Utilities v Australian National Airlines Comm. (1963) 80 WN (NSW) 48

Heydon's Case (1584) 3 Co Rep 7a at 7b; 76 ER 637.

Macquarie Bank v. Berg [1999] NSWSC 526

Tosier and Wife v Hawkins (1885) 15 QBD 680.

Voth v Manilda Flour Mills Pvt. Ltd (1990) 171 CLR 538, 565

XYZ v Commonwealth (2005) 227 ALR 495

References

Ackerman, M., Hofer, E. C., & Hanisch, R. (2008). The National Virtual Observatory. In G. Olson, A. Zimmerman & N. Bos (Eds.), *Scientific Collaboration on the Internet* (pp. 135–142). Cambridge, Massachusetts: The MIT Press.

ACLS. (2009). What is cyberinfrastructure? Retrieved October 25, 2009, from http://www.acls.org/programs/Default.aspx?id=644

Arzberger, P., Schroeder, P., Beaulieu, A., Bowker, G., Casey, K., Laaksonen, L., et al. (2004). An international framework to promote access to data. *Science, 303*(5665), 1777–1778.

Atkins, D. E., Droegemeier, K. K., Feldman, S. I., Garcia-Molina, H., Klein, M. L., & Messina, P. (2003). Revolutionizing science and engineering through cyberinfrastructure: Report of the National Science Foundation blue-ribbon advisory panel on cyberinfrastructure. Washington, DC: National Science Foundation. Retrieved December 19, 2006 from *http://www.communitytechnology.org/nsf_ci_report/*.

Baker, K. S., Ribes, D., Millerand, F., & Bowker, G. C. (2005). *Interoperability Strategies for Scientific Cyberinfrastructure: Research Practice*. Paper presented at the American Society for Information Systems and Technology.

Balkin, J. M. (2006, April 27). The democratic case for network neutrality. Retrieved October 15, 2009, from http://balkin.blogspot.com/2006/04/democratic-case-for-network-neutrality.html

Beardsley, S., & Farrell, D. (2005). Regulation that's good for competition. *The McKinsey Quartley, 2.*

Benkler, Y. (2006). *The wealth of networks: How social production transforms markets and freedom.* New Haven: Yale University Press.

Benvenisti, E. (2008). The Conception of International Law as a Legal System. *Tel Aviv University Law Faculty Papers.* Tel Aviv University Law School.

Berner-Lee, T. (2006, June 21, 2006). Net Neutrality: This Is Serious, Retrieved June 21, 2006, from http://dig.csail.mit.edu/breadcrumbs/node/144

Bird, I., Jones, B., & Kee, K. (2009). The organization and management of grid infrastructures. *Computer, 42*(1), 36–46.

Borgman, C., Wallis, J., Mayernik, M., & Pepe, A. (2007). *international and Interdisciplinary collaboration in cyberinfrastructure: A case study with embedded networked sensor technology.* Paper presented at the AAAS Annual Meeting. Presentation retrieved from

Boulle, L. (1996). *Mediation: Principles, Process, Practice.* Butterworths: Tottel Publishing.

Bresnan, T. F., & Greenstein, S. (1999). Technological Competition and the Structure of the Computer Industry. *The Journal of Industrial Economics, 47.*

Buckingham, D. (2007). Digital Media Literacies: Rethinking Media Education in the Age of the Internet. *Research in Comparative and International Education, 2*(1), 45.

Burk, D. L. (2007). Intellectual Property and Cyberinfrastructure. *First Monday, 12*(6).

Burmeister, K. (1999). Jurisdiction, Choice of Law, Copyright, and the Internet: Protection Against Framing in an International Setting. *Media & Entertainment LJ., 9*, 625.

Cave, M., & Crocioni, P. (2007). Does Europe Need Network Neturality Rules? *International Journal of Communication, 1*, 669–679.

CEDA, C. f. P. D. (2006). Reclaiming our Commonwealth: Policies for a Fair and Sustainable Future. *Common Sense Paper, 1.*

Cerf, V. (2006). Prepared Statement of Vinton G. Cerf to U.S. Senate Committee on Commerce, Science, and Transportation Hearing on "Network Neutrality."

Cole, R. J., & Lorch, R. (Eds.). (2003). *Buildings, culture & environment: Informing local & global practices.* Oxford: Blackwell Publishing Ltd.

Community Legal Framework for a European Research Infrastructure Consortium (ERIC), Council Regulation (EC) No 723/2009 C.F.R. (2009).

Contractor, N. (2009). The emergence of Multidimensional Networks. *Journal of Computer-Mediated Communication, 14*(3), 743–747.

Coughlan, S., Currie, R., Kindred, H., & Scassa, T. (2006). Global Reach, Local Grasp: Constructing Extraterritorial Jurisdiction in the Age of Globalization Law Commission of Canada.

Crandall, R., & Jackson, C. (2001). *The $500 Billion opportunity: The potential economic benefit of widespread diffusion of broadband internet access.* Washington, D.C., : Criterion Economics.

Cranston, R. (1986). What Do Courts Do? *Civil Justice Q, 5*(124).

Crump, B., & McIlroy, A. (2003). The Digital Divide: Why the "don't-want-tos" Won't Compute: Lessons from a New Zealand ICT Project.' *First Monday, 8*(12).

Cyberinfrastructure Research Taskforce. (2005). Final Report of the Indiana University Cyberinfrastructure Research Taskforce: Indiana University.

David, P., & Spence, M. (2003). Towards Institutional Infrastructures for E-Science: The Scope of the Challenge: Oxford Internet Institute.

David, P., & Spence, M. (2008). Designing Institutional Infrastructures for e-Science. In B. Fitzgerald (Ed.), *Legal Framework for E-Research: Realising the Potential.* Sydney: Sydney University Press

Department of Broadband, C. a. D. E. D. (2008). Annual Report 2007–2008: Australian Government.
Depypere, S. (1995). Speech—Why do we a need a competition policy? : Europa.
Dierickx, S. (2006). Web Analytics: What about Packet Sniffing? Retrieved May 31, 2006, from http://webanalytics.ox2.eu/2006/05/31/web-analytics-what-about-packet-sniffing/
Eccles, K., Schroeder, R., Meyer, E. T., Kertcher, Z., Barjak, F., Huesing, T., et al. (2009, 24–26 June). *The Future of E-Research Infrastructures.* Paper presented at the the 5th International Conference on e-Social Science (Proceedings), Cologne, Germany.
Editorial, A. T. (2009). Neutrality Vital to Health of Internet *St. Petersburg Times* Retrieved September 23, 2009, from http://www.tampabay.com/opinion/editorials/article1038353.ece
Edwards, M. (2001). *Social Policy, Public Policy: From problem to practice.* Crows Nest, NSW: Allen & Unwin.
Edwards, P., Jackson, S., Bowker, G., & Williams, R. (2009). Introduction: An agenda for infrastructure studies. *Journal of the Association for Information Systems, 10*(5), 364–374.
Endres, J. (2009). Net neutrality—How relevant is it to Australia? *Telecommunications Journal of Australia,, 59*(2), 22.21—22.10.
Fitzgerald, B., & Pappalardo, K. (2008). The Law as Cyber Infrastructure. In B. Fitzgerald (Ed.), *Legal Framework for E-Research: Realising the Potential.* Sydney: Sydney University Press
Foster, I., Kesselman, C., & Tuecke, S. (2001). The Anatomy of the Grid: Enabling Scalable Virtual Organizations. *International Journal of High Performance Computing Applications, 15*(3), 200–222. doi: 10.1177/109434200101500302
Friendlander, A. (2008). The Triple Helix: Cyberinfrastructure, Scholarly Communication, and Trust. *Journal of Electronic Publishing, 11*(1), Retrieved on June 13, 2008 from http://quod.lib.umich.edu/cgi/t/text/textidx?c=jep;view=text;rgn=main;idno=3336451.3330011.3336109.
Fry, J., & Schroeder, R. (2009). Towards a sociology of e-research: Shaping practice and advancing knowledge. In N. W. Jankowski (Ed.), *e-Research: Transformation in Scholarly Practice* (pp. 35–53). New York: Routledge.
Gibbs, S. H. (2000). Oration Delivered at the Opening of the Supreme Court Library's Rare Books Room at the Supreme Court of Queensland.
Goldfinch, S., Gauld, R., & Herbison, P. (2009). The Participation Divide? Political Participation, Trust in Government, and E-government in Australia and New Zealand. *Australasian Journal of Public Administration, 68*(3), 333–350.
Gunkel, D. (2003). Second Thoughts Toward a Critique of the Digital Divide.' *New Media & Society, 5*(4), 499–522.
Hai, Z. (2004). China's E-science Knowledge Grid Environment. *Intelligent Systems, IEEE, 19*(1), 13–17.
Hammond, A. S. (2002). The Digital Divide in the New Millennium. *20 Cardozo Arts & Entertainment L.J,* 135–156.
Hofer, E. C., McKee, S., Birnholtz, J. P., & Avery, P. (2008). High Energy Physics: The Large Hadron Collider Collaborations. In G. M. Olson, A. Zimmerman & N. Bos (Eds.), *Scientific Collaboration on the Internet* (pp. 143—151). Cambridge, MA: MIT Press.
Holland, T. E. (2006). *The Element of Jurisprudence.* Clark, New Jersey: The Lawbook Exchange Ltd.
Jankowski, N. W. (2007). Exploring e-science: An introduction. *Journal of Computer-Mediated Communication, 12*(2), article 10. http://jcmc.indiana.edu/vol12/issue12/janakowski.html.
Jankowski, N. W. (2009). *E-Research: Transformation in Scholarly Practice.* New York, NY: Routledge.
Johnson, K. (2009). The importance of net neutrality to the digital economy. *Telecommunications Journal of Australia,, 59*(2), 19.11.

Kariyawasam, R. (2007). *International Economic Law and the Digital Divide: A New Silk Road* Cheltenham: Edward Elgar Publishing Limited.

La Rose, R., Gregg, J., Strover, S., Straubhauer, J., & Inagaki, N. (2007). Closing the Rural Broadband Gap: Promoting Adoption of the Internet in Rural America. *Telecommunications Policy, 31*, 359–373.

Latour, B. (2005). *Reassembling the social: An introduction to actor-network-theory.* Oxford: Oxford University Press.

Lee, C., & Tibbo, H. (2007). Digital Curation and Trusted Repositories: Steps Toward Success. *Journal of Digital Information, 8*(2).

Lee, C. P., Dourish, P., & Mark, G. (2006). The human infrastructure of cyberinfrastructure. In P. Hinds & D. Martin (Eds.), *CSCW '06: Proceedings of the Conference on Computer Supported Cooperative Work, Banff, Alberta, Canada* (pp. 483–492). New York: ACM Press.

Leonardi, P. (2009). Why Do People Reject New Technologies and Stymie Organizational Change of Which They Are in Favor? Exploring Misalignments between Social Interactions and Materiality. *Human Communication Research, 35*, 407–441.

Lessig, L. (1999). The Law of the Horse: What Cyber Law Might Teach. *Harvard Law Review.*

Lessig, L. (2009). Code version 02. Retrieved December 14, 2009, from http://codev2.cc/download+remix/

Lessig, L., & McChesney, R. (2006, June 8th). No Tolls on the Internet. *The Washington Post.* Retrieved from http://www.washingtonpost.com/wpdyn/content/article/2006/06/07/AR2006060702108_pf.html

Lynch, C. (2008). The Institutional Challenges of Cyberinfrastructure and E-Research. *EDUCAUSE Review, 43*(6).

Maldonado, E., & Tapia, A. (2007, September 28–30). *Government-Mandated Open Source Development: The Case Study of Venezuela.* Paper presented at the Telecommunication Policy Research Conference, Washington, DC, US.

Messigian, A. (2006). Love's Labour's Lost: Michael Lewis Clark's Constitutional Challenge of 18 U.S.C. 2423 (C). *The American Criminal Law Review, 43*, 1241.

Meyer, E. T., & Dutton, W. H. (2009). Top-down E-infrastructure Meets Bottom-up Research Innovation: The Social Shaping of E-research *Prometheus, 27*(3), 239–250.

Meyer, E. T., Schroeder, R., & Dutton, W. H. (2008, 28 February–1 March). *The role of e-infrastructures in the transformation of research practices and outcomes.* Paper presented at the iConference UCLA, Los Angeles, CA.

Michaels, R., & Jansen, N. (2007). Private Law Beyond the State? Europeanization, Globalization, Privatization. *Duke L School Working Paper Series,* . Duke L School Faculty Scholarship Series.

Middleton, G. (2008). e_Research and Jurisdiction. In B. Fitzgerald (Ed.), *Legal Framework for E-Research: Realising the Potential.* Sydney: Sydney University Press.

Mitchell, W. J. (2000). *E-topia.* Cambridge, MA: The MIT Press.

Monteiro, M., & Keating, E. (2009). Managing misunderstandings: The Role of Language in Interdisciplinary Scientific Collaboration. *Science Communication, 31*(1), 6–28.

Murray, A. D. (2007). *The Regulation of Cyberspace: Control in the Online Environment.* Oxon: Routledge-Cavendish.

Myer, J. (2008). A National User Facility That Fits on Your Desk: The Evolution of Collaboratories at the Pacific Northwest National Labratory. In G. Olson, A. Zimmerman & N. Bos (Eds.), *Scientific Collaboration on the Internet* (pp. 121–134). Cambridge, Massachusetts: The MIT Press.

Nakamura, L. (2000). Economics and the New Economy: The Invisible Hand Meets Creative

Destruction. *Business Review, July/August.*
National Science Foundation Cyberinfrastructure Council. (2007). *Cyberinfrastructure vision for 21st century discovery.* Arlington, VA: Retrieved from http://www.nsf.gov/pubs/2007/nsf0728/nsf0728.pdf.Olson, G., Zimmerman, A., & Bos, N. (Eds.). (2008). *Scientific collaboration on the internet.* Cambridge: MIT Press.
Olson, J., Ellisman, M., James, M., Grethe, J., & Puetz, M. (2008). The biomedical informatics research network. In G. M. Olson, A. Zimmerman & N. Bos (Eds.), *Scientific collaboration on the Internet* (pp. 221—232). Cambridge, MA: MIT Press.
Petrazzini, B., & Kibati, M. (1999). The Internet in Developing Countries. *Communications of the ACM, 42*(6), 31–36.
Poole, M. S. (2009). Collaboration, Integration, and Transformation: Directions for research on communication and information technologies. *Journal of Computer-Mediated Communication, 14*(3), 758–763.
Prensky, M. (2001). Digital Natives, Digital Immigrants Part 1. *On the Horizon, 9*(5), 1—6.
Reid, T. A. (2007, May 2). *The new holy grail: An Australian e-infrastructure.* Paper presented at the EDUCAUSE Australasia, Melbourne, Australia. http://www.caudit.edu.au/educauseaustralasia07/authors_papers/Reid-238.pdf.
Ribes, D., & Finholt, T. A. (2009). The Long Now of Technology Infrastructure: Articulating Tensions in Development. *Journal of the Association for Information Systems, 10*(5), 375–398.
savetheinternet.com. (2009). FAQ Retrieved November 8 2009, from http://www.savetheinterent.com
Schroeder, R. (2007a). e-Research Infrastructures and Open Science: Towards a New System of Knowledge Production? . *Prometheus, 25*(1), 1–17.
Schroeder, R. (2007b). *Rethinking science, technology, and social change.* Stanford, CA: Stanford University Press.
Schroeder, R., & Axelsson, A. (Eds.). (2006). *Avatars at work and play: Collaboration and interaction in shared virtual environments.* Dordrecht: Springer.
Seidel, E., Muñoz, J., Meacham, S., & Whitson, C. A. (2009). A Vision for Cyberinfrastructure. *Computer, 42*(1), 40.
Stewart, C. (2007). Indiana University cyberinfrastructure newsletter. Retrieved November 1, 2008, from http://racinfo.indiana.edu/newsletter/archives/2007-03.shtml
Tapia, A., Blodgett, B., & Jang, J. (2009). *The merging of telecommuncations policy and science policy through broadband stimulus funding.* Paper presented at the TPRC Research Conference on Communication, Information and Internet Policy, Arlington, VA.
Unsworth, J. (2008). Cyber Infrastructure for the Humanities and Social Sciences. In B. Fitzgerald (Ed.), *Legal Framework for E-Research: Realising the Potential.* Sydney: Sydney University Press.
White, L. J. (2008). *The Role of Competition Policy in the Promotion of Economic Growth.* New York University School of Law and Economics Working Papers. New York University School of Law.
Wu, T. (2003). Network Neutrality FAQ Retrieved November 8 2009, from http://www.timwu.org/network_neutrality.html
Ypsilanti, T., & Paltridge, S. (2004). OECD Broadband Market Developments. In R. Cooper & G. Madden (Eds.), *Frontiers of Broadband, Electronic and Mobile Commerce.* Heidelbery: Physcia-Verlag.
Zittrain, J. (2006). The Generative Internet. *Harvard Law Journal, 119.*

SECTION THREE

Political Intersections

CHAPTER NINE

Leetocracy [1]

Networked Political Activism or the Continuation of Elitism in Competitive Democracy

YANA BREINDL & NILS GUSTAFSSON

INTRODUCTION

On May 6, 2009, the European Parliament (EP) gathered for its monthly plenary in Strasbourg. Among the texts to be voted upon was the so-called Telecoms Package, a set of five directives regulating the European telecommunications market, which the three major institutions of the European Union (EU)—the European Commission, the European Parliament, and the Council of the EU—have been working on for nearly 2 years. The outcome of the vote seemed settled, as the Council and the EP had come to an agreement on the entire package and were hoping to close it before the European elections of June 2009. However, the vote did not follow the initially assumed voting order. Instead of confirming the compromise previously agreed with the Council, the members of the EP (MEPs)—to the surprise of many observers—adopted the initial version of "amendement 138," postponing the entire negotiations for a further 6 months as the EP and Council extended the negotiations to a third reading. Amendement 138 states that member states cannot cut off internet access without a prior ruling of the judicial authorities, and is strongly supported by a French advocacy group, la Quadrature du Net (Squaring the Net), that constituted itself in March 2008 to try to prevent French and EU legislators from passing repressive laws on the digital realm. How can a group, referred to as "five blokes in a garage" by a senior French civil servant,[2] introduce such an amendment to a highly complex package and effectively lobby decision makers, thus reversing a set deal at the last minute?

This chapter critically examines the role of networked advocacy groups in the policy-making process of intellectual property rights reform. Through analysing the case of la Quadrature du Net, we question the assumption that political intermediaries or elites are disappearing leaving space for a more inclusive, direct democracy in which decision makers interact more directly with citizens. Next to established political actors such as political parties or trade unions, which are— sometimes reluctantly—integrating information and communication technologies (ICTs) into their working practices, internet-based actors are emerging in a wide range of political areas (Chadwick, 2006). Such forms of networked political organisations are usually perceived as less hierarchical than traditional mobilizing groups (Norris, 2002; Dalton, 2008). This development is often interpreted by techno-optimists as a way out of the iron law of oligarchy in traditional politics, offsetting the professionalization of politics and the transfer of political power to technocrats and anonymous international political actors far away from democratic accountability, thus preparing the ground for a more inclusive grassroots-oriented democracy.

However, we argue that intermediary elites still exist. After a short discussion of the articulation between ICTs and democratic theory ("Democracy in a Digital Age" section), we introduce the concept of temporal elites ("Temporal Elites" section) and apply it to the case of la Quadrature du Net's campaign surrounding the Telecoms Package. This campaign has proven successful in the sense that it has had a clear impact upon the decision-making process ("Internet-Based Lobbying on the Telecoms Package" section). Our discussion will show that internet-based activism constitutes a new type of elites in competitive democracy, whose effective forms are heavily dependent on technical and networking skills ("Discussion" section). Rather than functioning as the base of more egalitarian politics, the growing importance of networked political activism aided by digital media may, on the contrary, create new elites. We finish by discussing whether such elites are detrimental or beneficial to a well-functioning democracy.

DEMOCRACY IN A DIGITAL AGE

Contemporary democratic politics is characterized by the uneasy coexistence of old power networks sustained by elites in the parliamentary political system with, arguably, two types of actors: an intermediary level of business elites, mass media outlets, and interest groups on the one hand; and newer forms of organized interests, functioning in seemingly unstable, unprecedented, and unpredictable networks of individuals and organizations, generally associated with and aided by

ICTs, on the other (cf. Norris, 2002; Dalton, 2008; Wellman, Haase, Witte, & Hampton, 2001; Castells, 1996; Micheletti, 2003; Shirky, 2008). This development is connected to several central debates that have arisen in the social sciences in the past decades, including discussions about the rise of postmaterialist values in (post-)industrialized countries (Inglehart, 1977), and the growth of new forms of political activism (Micheletti, 2003; Bentivegna, 2006; Baringshorst, 2009). This kind of renewed participation supposedly takes place beyond conventional forms of political participation and is known as "life politics" (Giddens, 1991) or "sub politics" (Beck, 1997). As traditional political institutions become increasingly contested in an era of globalization and corporatism characterized by dense networks of communication, politics are "materializing in different ambits and contexts, thus meaning the loss of 'center' as a consequence of the crumbling of the traditional political institutions that previously had control of it." (Bentivegna, 2006, p. 332).

The waning of social capital and political participation as a result of individualization (Putnam, 2000), or conversely, social capital being reinforced by individualization, resulting in new forms of participation (Dalton, 2008; Dahlgren, 2009), is equally subject to debate. As ICTs have become more prevalent, cheaper, and useful since the rapid spread of internet connections in the 1990s, social science has increasingly turned its eyes towards the web as a promise of a more democratic future (e.g. Rheingold, 2000; Becker & Slaton, 2000; Morris & Delafon, 2002; Lévy, 1997, 2002) or as a dynamic machine concentrating ever more power into the hands of the few (e.g. Van de Donk, Snellen, & Tops, 1995; Hindman, 2008). The development of applications often referred to as Web 2.0 and social media in the mid-2000s, combined with anecdotal evidence of new forms of rapid networked mobilization (cf. Rheingold, 2002; Jenkins, 2006; Benkler, 2006; Shirky, 2008), created a new interest in the effects of technology on political participation.

Much of the literature on the articulation between politics and ICTs is underpinned by a certain dissatisfaction with the electoral democratic system prevalent in many industrialized nations (Norris, 2002). Current political systems are frequently considered as failing to fulfil ideals emphasizing egalitarianism, affecting not only formal political rights like voting, freedom of speech, freedom to organize and so on, but also who actually participates in political life, sets the agenda and makes the decisions. Like the invention of previous technologies such as the telegraph, the radio, or television (Vanobberghen, 2007; Hoff & Bjerke, 2009), the internet fostered hopes for an invigorated public using technologies to learn about and promote political and social causes for the good of humanity. Political participation makes people grow as individuals, leading to emancipation as well as to better governance (Norris, 2002, p.5). Hence, the debate has centered on the need for mass participation and whether internet use promotes it or not. These sets of democratic ideals or theories are summarized by deliberative democratic theory, emphasizing rational political discourse in the public sphere (Habermas, 1962/1989, 1996; Fishkin, 1991; Barber, 1984) and direct democracy, emphasizing the actual participation in politics by all (or a large number of) citizens (Pateman, 1970 ; Beitz, 1989). It is not difficult to understand how this has come about; the history of the internet is also a history of a libertarian, anti-establishment, meritocratic, and anti-hierarchical culture (Castells, 2001; Rasmussen, 2007). From a communication science perspective, the internet enables not only traditional one-to-many communications, but also the possibility for many-to-many conversations. It allows communication with as many people as one wishes, providing one has access to the network. In terms of political activism, the mass coordination of large groups of people becomes possible.

The existing, imperfect system of electoral democracy can also be associated with what David Held (2006) calls competitive elitism. Competitive elitism, which has been laid out in its purest theoretical form by Joseph Schumpeter (1946) and Anthony Downs (1957), employs the economic model of rational action in markets to analyze the democratic system. In essence, the democratic model, presented as both existing and desirable in these treatises, assigns citizens the passive role of voters, selecting their rulers among competing elites. The emphasis is put on fostering competent politicians, experts in their fields (which the public cannot be), but accountable for their actions to the electorate (competition assures quality in governance).

Although the above-mentioned normative theories of what democracy is (or should be) do not completely describe any existing political system, and although elements of the theories tend to coexist in reality, we nevertheless believe that the competitive elitist model resembles the state of contemporary democracy more than do the deliberative and direct democratic models. One aim of this chapter is to turn to the members of the political elites themselves, rather than trying to analyze movements and campaigns aided by digital tools as examples of increased deliberation or mass participation.

The core problem with understanding the current political reality is the failure to see that internet-based networks do not pose a threat to the competitive elitist democratic system of our time, nor are they simply a continuation of old structures. Instead, they represent a complementary tool of informing the political elite about the wishes of certain parts of the electorate. In the competitive elitist model there is no dual model where society consists of powerful politicians/rulers and voters/ruled with extremely limited power. Instead, the political system of a society can be analyzed as a series of strata with the key decision makers at the top, a large group of fairly passive bystanders who restrict their actions to voting, a smaller group of nonvoters completely disinterested in the political games, and opinion leaders and activists acting as intermediaries between these strata.

Temporal Elites

David Miller (1983, p.134) describes the political elite as: "a small group of political leaders, [...] with perhaps an intermediate section of more active citizens, who transmit demands and information between the mass and the leadership." This intermediary group of influentials and activists as described by Putnam (1976, see Figure 1) can be further divided into various strata. The actual power exerted by this group of people is directed both "up" and "down": activists influence politicians directly as well as the "mass," who in turn exert influence over the politicians. Whereas some supporters of the direct or deliberative democratic model claim that digital tools might render such intermediaries obsolete as direct contact between leaders and citizens is made possible, we argue that core activists form a new elite, augmenting the existing model.

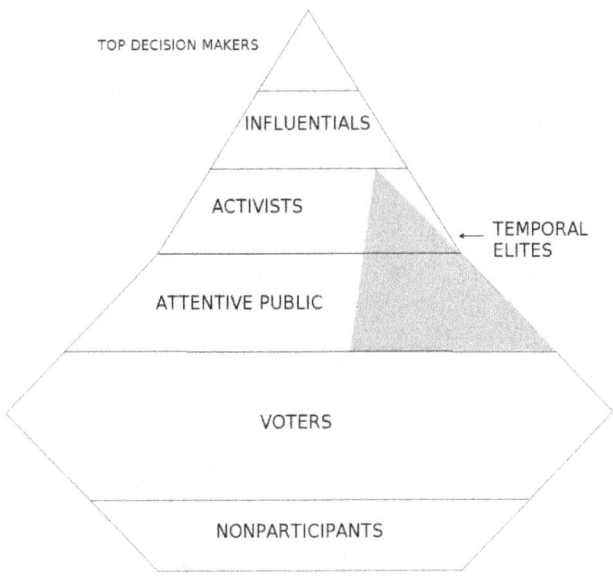

Figure 1. Putnam's pyramid of power (adapted with modifications from Putnam, 1976)

1. *Top decision makers*: incumbents in key official posts. This is normally a very small group of people.
2. *Influentials*: powerful opinion makers and people to whom decision makers look for advice—high-level bureaucrats, interest group leaders. This is also a small group.
3. *Activists*: This stratum is made up of the group of citizens who take active part in politics—as members of a political party or on a more private level. This is a larger group of people.
4. Still larger is the stratum of the *attentive public*, which consists of citizens who follow the political debates as some kind of spectator sport. They rarely actively participate.
5. The main bulk of citizens are the *voters* who have very limited, if any, political influence. They vote and that is all.
6. Finally, the *nonparticipants* do not even vote and have no political power regarding the formal political system.

We refer to networked activists as *temporal elites* (Gustafsson, 2009). The concept denotes their limited influence on certain fields and their highly unpredictable success in exerting influence over policy outcomes and agenda-setting. In terms of Putnam's model, temporal elites belong to the third and fourth strata (as shown earlier), with fairly inactive supporters of the campaign belonging to the fourth, and the attentive public and core of highly involved individuals to the third strata, the activists. However, we suppose that the group of activists grows in importance compared to the second stratum, the influentials, as "viral politics" rise in importance as compared to traditional means of influencing politicians and the public. Temporal elites adopt "viral politics," or the rapid sharing of information across the internet resulting in political mobilization (Gustafsson, 2009).

The strategy behind viral politics is to increase the number of persons composing the fourth strata, the attentive public, paying attention to the specific campaign. These "spectators" may not contribute actively to the campaign, but the more interest there is for an issue, the more politicians feel they are watched and are likely to listen to activists. Having no direct access to the mainstream media, temporal elites effectively use ICTs to spread their messages on a multiplicity of platforms. In this sense, they truly challenge established actors, although the success of a viral campaign is often measured by the resonance it creates in traditional media outlets. Furthermore, temporal elites work towards convincing "spectators" to make the step to the third strata, the activists. At the same time, activists try hard to establish themselves as influentials or to turn decision makers or influentials into activists. This form of shifting strata is not radically new. On the contrary, it is characteristic of any type of contestation, which by definition aims to alter existing power structures. What distinguishes internet-based activism from previous forms of campaigning is the reach their message can potentially have at a relatively low cost and the loose organizing forms such activism adopts.

In specific networked political campaigns, we usually find that the "movement" mobilizing around the issue can be imagined as a number of concentric circles, with a core of dedicated activists in the middle that we can call political or movement entrepreneurs (Gustafsson, 2009). (Note that the "pyramid of power" graphically represents the power strata in all society and not the power balance in specific campaigns or movements). Such individuals are generally directly affected by the issue at stake, and rely upon their own skills to achieve their objectives (Earl & Schussman, 2003; Gustafsson, 2009). Sometimes they act out of individual grievances (Earl & Schussman, 2003), but are generally nodes in a larger network of activists who share common views and notions of political strategy (Gustafsson, 2009). Movement entrepreneurs active on different levels or countries do not necessarily know each other personally but observe each other on the internet, developing a common understanding of a certain political issue (Baringhorst, 2009). Core campaigners often spend uncountable hours on the campaign, frequently full time—at least during key moments of the campaign. From the core to the periphery we can then see circles containing first activists who spend large amounts of time volunteering for the campaign, then people who contribute only occasionally, and finally, a wide, shifting group of "lurkers" who intermittently participate through informing themselves about the issue. The core and other activists belong in Putnam's model among the activists, whereas the lurkers belong to the attentive public.

The people closer to the core can be described as more powerful than the ones in the periphery due to their often more detailed knowledge of the issue. Yet, they are usually powerless without a wider supporting group who can spread information through social networks and rapidly mobilize. They often possess the features we usually associate with political influence: education, technical skills, sociability, and organizational skills, but they are nonetheless also an example of a disruptive force in the existing elitarian system. They do not necessarily need large financial interests behind them, nor massive organizations with thousands of card-carrying members, willing to make phone calls and sit in tedious meetings on weeknights. They benefit from the way ICTs enable "flexible participation" (Joyce, 2007); the barrier of entry into political activism is lowered by the fact that the repertoire of actions and the time and resources needed to participate in a campaign can be individualized to fit every participant's schedule and interest. One of the key advantages the internet offers is that it allows the efficient aggregation of small contributions—for example, in terms of time spent on sharing information, donating money, editing the wiki, contributing to the planning of a mobilization, or contacting an MEP. The next section will examine the case of la Quadrature du Net and its Telecoms Package campaign, before developing the concept of temporal elites and viral politics in light of this case.

INTERNET-BASED LOBBYING ON THE TELECOMS PACKAGE

La Quadrature du Net (QdN) is a French citizen-collective established in March 2008 in response to president Nicolas Sarkozy's announcement of the introduction of a three-strikes plan, negotiated with the record industry and internet providers. Their name refers to the impossible mathematical problem of "squaring a circle," an analogy for the impossibility of transposing traditional legislation onto the digital environment. QdN believes that "it is impossible to effectively control the flow of information in the digital age by means of the law and technology without harming public freedoms, and damaging economic and social development. This is what we call Squaring the Net."[3] The advocacy group therefore calls for innovative internet regulation that respects fundamental rights and the inherently democratic character of the internet.

The data for this research was collected as part of a broader project that analyses campaigns aiming to influence EU policy-making in the domains of intellectual property rights, internet regulation and so-called digital rights, that is, the protection of citizens rights in digital realms (cf. Breindl & Briatte, 2009; Breindl, 2010). The Telecoms Package campaign has been selected because it is a typical example of an internet-based network of activists campaigning to influence EU decision-making on copyright issues. Activism surrounding intellectual property rights lends itself heavily to internet activism, as the tools and objectives they pursue coincide. QdN is emblematic of networked, transnational, internet-based activism in a domain that heavily relies on ICTs.[4]

QdN is a hybrid organisation (Chadwick, 2006), mixing the action repertoires traditionally associated with social movements and interest groups: protest actions (such as an internet blackout), but also participation in conferences, discussions with MEPs, and the provision of analyses. QdN is part of an international network of digital rights advocates. Their aim is to prevent what they consider repressive copyright legislations such as the "three-strikes-and-you're-out" scheme that plans to cut off copyright infringers' internet connection after two unsuccessful warnings. Amendement 138 was introduced as a warrant against such a scheme, as it would make a prior judicial ruling compulsory, complicating the three-strikes mechanism. Three-strikes is only the latest in a series of events generally referred to as the "copyright wars" in which the entertainment industry uses any possible venue in order to counter copyright infringements, including lobbying, litigation, education, and licensing (Yu, 2004).

Internet-based networks such as QdN have been instrumental in raising public awareness of copyright issues (Breindl & Briatte, 2009), and are typical of a larger trend of new communities that have emerged with the rapid expansion of the internet. Since the early 1990s, they are gaining in importance, notably by influencing traditional decision-making. On their internet site, they define their activities as advocacy "for the adaptation of French and European legislations to respect the founding principles of the Internet, most notably the free circulation of knowledge," and intervening in "public-policy debates concerning, for instance, freedom of speech, copyright, regulation of telecommunications and online privacy."[5] More generally, their actions aim to encourage citizen participation and debate on "rights and freedoms in the digital age.[6]"

Most core activists can to some extent be linked to the free/libre and open source software movement (FLOSS), either as programmers, free or open source software company owners or users. For these activists, the advent of computers and the internet is a revolution that fundamentally alters the current power balance, moving from an industrial society to an information society. They are inspired by what Castells has termed the "culture of the internet" (2001, pp. 36–63), based on the techno-meritocratic values built in the open architecture of the internet by its early innovators; enacted by hackers promoting principles of sharing, openness, decentralization, free access to computers and information, and the belief that computers can change the world for the better (Levy, 1984); and embedded in virtual communitarian networks and the entrepreneurial culture that contributes to "an ideology of freedom that is widespread on the Internet world" (Castells 2001, p.37; see also Flichy, 2001; Rasmussen, 2007). At present, a much broader digital rights movement has taken shape, as exemplified by QdN's promotion of openness, sharing, and free access.

The frames articulated by digital rights activists are notable for their trans-political appeal, resisting traditional right/left cleavages. QdN succeeded in playing on antagonisms within the two big European political formations, the European People's Party (EPP) and the Progressive Alliance of Socialists and Democrats in the EP (S&D), leading to affinities between members of different parties and a crucial role for small parties to act as intermediaries. This is, however, not unusual in European politics, characterised by shifting majorities depending on the issues at stake. From QdN's perspective, governments and corporations frequently do not understand the emancipatory potential of internet technologies and try to regulate them in order to control them more effectively. In a European Parliament largely dominated by the conservative EPP, their sole chance of success is to increase the awareness of MEPs across the political spectrum.

The way they work reveals a strong tendency to adopt an "engineering philosophy to make things work" and an "insistence on adopting a technocratic approach to solving societal problems and to bypassing (*hacking*) legislative approaches" (Berry 2008, p.102). If there are "harmful" amendments within a French legislative proposal or even within a set of five European directives, everything needs to be done to "patch"[7] these, as one activist explains:

> Basically, what you had in this kind of community is a certain pragmatic approach towards implementing stuff, by doing stuff and problem solving. So you have a problem, try to get a fix for it, try to get a solution. You're not so much interested as other political communities in socializing or in feeling good among us and sticking together as a community. So this doesn't really matter. We want to achieve our objective. Yeah. It's very focused. (...) Actually, politics is also a technocratic system and in the same way you program computers, you somehow try to fix the political regulatory framework. (Interview 1, Brussels, February 2008).

QdN can be best described as functioning in four concentric circles, as introduced previously in the discussion on temporal elites. At its core are five founders, four of whom are computer scientists, empathetic to the FLOSS movement, and the fifth previously a parliamentary assistant in the French national assembly before rallying to the digital rights cause. One core campaigner and a half-time assistant are paid with funds provided by the Open Society Institute (OSI). Founded by the Hungarian-American businessman and philantropist George Soros, OIS is a private foundation offering grants for the promotion of democratic governance and the safeguard of fundamental rights. The second circle is composed of voluntary contributors who are generally part of la Quadrature du Net's discussion list and follow the Internet Relay Chats (IRC). These contributors do not only actively engaging in the discussion but analyze legislative texts, check press releases, edit the campaign wiki, create the word online, and create new tools. A third circle is composed of occasional contributors, people who follow closely what la Quadrature du Net does, performing tasks such as translating documents or the content of the website, cleaning up the wiki or helping out with reviewing the press coverage of their activities. Finally, a fourth circle of supporters, so-called "lurkers," is comprised of people who read and follow what la Quadrature du Net does, maybe engaging in their mobilizational campaigns through calling an MEP or participating in the internet blackout,[8] but without actively contributing to the organisation of the campaign itself.

Most (but not all) core campaigners and supporters interviewed are male, holding a university degree, aged between 20 and 35 and live in urban areas. The boundaries between these circles are far from impermeable. Even core activists can put their activities on hold for a certain period of time and become occasional contributors, just as lurkers can decide to join the IRC discussions and move closer to the core of the group. These dynamics are oberservable within online groups in general. Often a core group of very active members is responsible for most of the produced content, while up to 90% are made up of lurkers (Nonnecke & Preece, 2000). Due to "[t]he fluid character of many of these netbased movements, and the ease of joining and withdrawing, it is really difficult to estimate what portion of the citizenry is actually involved" (Dahlgren, 2004, p.18).

QdN is an informal organisation, without statutes or an elected board. The collective emerged in response to the so-called HADOPI law[9] in France. Most core activists used to fight previous intellectual property rights legislations, such as the directive on computer implemented innovations or the copyright law DADVSI in France (Breindl & Briatte, 2009). These past struggles prompted their awareness of the necessity to look at the European level, if only because two-thirds of all legislations in member states legislations are transpositions from EU law.

In spring 2008 they discovered that among more than 700 amendments to the reform of the Telecoms Package (a set of five directives regulating the European telecommunications market), several weresupportive of establishing "graduated response" or "three strikes" legislation being established at a European level. Further amendments were problematic to the principle of net neutrality (the undiscriminated routing of content over the internet) or to the respect of privacy in digital realms. Their actions therefore became twofold; at the French level with the HADOPI law, and at the European level with the Telecoms Package.

The Telecoms Package campaign lasted over a period of about 20 months, closely following the legislative process through which QdN published numerous press releases. QdN actively worked to form alliances with like-minded activists and associations of other member states who would relay their message during the various mobilizations. Creating a network of involved individuals was a central component of the campaign and one for which the use of the internet is generally lauded by scholars (Castells, 2001; Bennett, 2004). However, the actors that held central positions within the network were not random citizens. On the contrary, the form of activism practised by QdN involves highly skilled actors. Most of them hold a university degree—frequently computer science but not exclusively—and as one ally inside the EP stated: "they generally come from privileged social classes or at least they have learnt everything that is necessary." (Interview 2, Brussels, March 2009). The internet does not remove all barriers to participation; education and social capital remain strong determinants of online action (Jensen, 2006).

Furthermore, they are not only privileged and intelligent individuals, they are also technically skilled, that is, they know how computers function, how the internet works, and in which way they can take advantage of these technologies by developing viral campaigns. For example, a benevolent founder of Quadrature du Net generated the tool LawTracks,[10] thanks to which any internet user can compare different versions of problematic articles of the Telecoms Package. A link to the software used for generating this database explains furthermore how it can be installed and adapted—freely—by other activists/associations. The original texts of the directive are extracted from EUR-Lex, a European platform that provides free access to EU law texts.[11] These texts are available in the official EU working languages (English, French, German, and Spanish) but further translations can be added.

Figure 2. Excerpt comparing the three institutions' changes to the Telecoms Package using LawTracks

The fact that la Quadrature du Net can rely on a large base of programmers certainly helps to build a coherent website and tools for analysis. Enabling citizen participation is a central component of QdN, with individuals asked to participate in various ways. They can contribute by looking at their wiki page "How to help" which lists the most recent tasks that need to be done. As such, it enables "flexible participation" (Joyce, 2007) even though most of the content is produced by the handful of core activists who rely on their technical expertise to build tools, such as LawTracks, that facilitate their intervention into EU policy-making. As one core campaigner asserts: "What I like most actually, it's to be a toolbox to allow people to understand what is happening and to allow them to act, to give them the tools to act." (Interview 12, Berlin, April 2009).

A recurrent claim of la Quadrature's press releases concerns the lack of transparency of policy-making. Decisions are taken in opaque committees and information is sometimes delivered to the public only once as changes can no longer be made. This is particularly the case regarding EU decision-making, which lacks strong mechanisms of democratic accountability. QdN's attempt to engage citizens with the complex EU system is particularly well received by political representatives who advocate the constitution of a strong European public sphere.

Discussion

QdN core campaigners can be described as temporal elites. They actively engage with politics, yet focus on a particular domain—internet regulation and intellectual property rights reform—using viral politics techniques to produce awareness and outreach. Thanks to their use of the web, they have not only aquired a good knowledge of a complex supranational policy system such as the EU, but have also used this expertise to take action and mobilize others to act. By continuously informing their readers via press releases, they try to involve citizens in the organisation of the campaign generating media resonance and/or putting pressure on political decision makers via phone calls and emails. On the internet, QdN has provided the most frequent updates on the Telecoms Package reform, from a politicized perspective, and their analyses have been widely read not only by their supporters, but also by their opponents. Their claims have frequently been relayed in the traditional media and across the EU as activists from other countries published and translated their releases.

Temporal elites are intermediaries between political decision makers and citizens, acting as transmitters of information from one section of the population to the other. Of course, not all Europeans have been touched by QdN's campaign, given that it is a very specialized domain. Yet, they managed to mobilize a significant portion of the citizenry, as all MEPs spoken to testified, regardless of their position on this issue. As QdN's prime goal is to influence existing representative democratic politics, they are not an alien element to the competitive elitist system. Instead, they manage to break inside the power pyramid previously described, effectively merging with the activists and the attentive public.

At the same time, the emergence of temporal elites does not mean the reinforcement of old elites. As barriers of entry are lowered and communication made easier, new groups formerly uninvolved in politics can be drawn in. However, as our case study shows, these new political participants have much in common with old elites with regards to social-economic-status (SES). Classical factors determining political participation such as time and money, education, social capital, and additional "digital factors" such as access, competency, motivation, and know-how, constitute barriers to participation (Jensen, 2006). Active minorities are often overrepresented in cyberspace (Corbineau & Barchechath, 2003). Hence, political actions, internet-based or not, are rarely representative or inclusive of the various groups constituting society. This is an important challenge to the principle of equality, central to all democratic models. The disruptive power of temporal elites and viral politics, instead, comes from the possibility of mobilising small groups of individuals around specified issues, thus competing with traditionally organized interests. The flexibility of participating in the campaign and the aggregation of small efforts allows for more people to become engaged.

The Telecoms Package campaign also shows how communication has become a primary political strategy, making "campaigns themselves political organizations that sustain activist networks in the absence of leadership by central organizations" (Bennett, 2004, p. 130). La Quadrature du Net constitutes a continuous campaign network, established to mobilise against a French law and soon moving to different levels. It is not an organization *stricto sensu* but an informal network of activists whose primary objective is to prevent "harmful" legislations within internet-related domains. Nonetheless, networks do not suppose that all of their members are equal, only that communication flows more horizontally—hierarchies are also networks. La Quadrature du Net and most contemporary forms of networked activism are indeed characterised by their interconnectedness and absence of strong leadership or central authority. However, within the various clusters composing networks such as QdN, some individuals hold more power than others, generally the most active ones.

QdN took advantage of the effective aggregation of small contributions and new forms of flexible participation. Yet, most of the work has been done by the small group of core campaigners who developed their expertise on internet-related issues. E-government practices have led to the publication of large amounts of official information on the internet. Even the European Union is keen on using these technologies to resolve the democratic deficit it is often accused of. "Netcitizens now dispose of research possibilities that used to only be accessible to State news services" argues Rebentisch (2005, p. 1). Yet, mere access to information does not necessarily increase participation levels. If the mass of information available is larger than before, it is not necessarily evenly spread. For this reason it requires increased expertise to find that information, and to understand, analyse and take advantage of it. This requires time, skills, and interest in engaging with such information, hence privileging some individuals over others. Groups such as la Quadrature du Net constitute new information gatekeepers, certainly working in favour of increased transparency in the political process, but still controlling what information is published as it relates to their cause.

Conclusion

In this chapter, to challenge the misconception that the emergence of new forms of digitally aided political activism, carried out through loose networks rather than through formal organisations, might be heralded as a positive replacement or at least a threat to the existing traditional elitist democractic systems of the world. Instead, we point to the way these new forms of organisation can produce new hierarchies and the emergence of new elites. We use the term temporal elites to describe a heterogeneous group of technologically and socially skilled activists with a strong motivation to influence policy, forming networks around specific issues with a few dedicated individuals in the core and larger groups of interested and potentially mobilizable people forming the important peripheral network. The term is useful for interpreting empirical studies of digital political activism in the light of elitist democratic theory, as our study of QdN shows. We do not claim that the evidence presented in this case is generalizable to all forms of protest activity relying on the internet nor that elitist democratic theory is the sole perspective through which to analyse what is happening in the field. Future research will have to address to what extent internet-based activism is disruptive for representative democracies and work on how to integrate various democratic theories and other conceptual frameworks to shed light upon the phenomenon.

In the end, whether temporal elites are seen as beneficial or detrimental to democracy is not only a question of democratic ideology, but also one of realism. Digital activism does not end elitism in democracy; it might on the contrary augment the existing system. But it is hard to claim that internet-based activists worsen the situation from an egalitarian point of view. Quite the contrary, the barriers for participation have been lowered. Motivated people with some basic skills can more easily than before use available information, build up a network of activists, and get the message into the political system (or out on the streets). Not everyone is motivated. Some people become interested in politics and, to make a long story short, we do not know why (Verba, Schlozman, & Brady, 1995). Furthermore, some people have the technical and social skills needed to participate successfully. This is connected to factors such as education and social background, but there is no evident unequivocal causality. We believe that political participation by as many as possible in a given society means better, more efficient, and more legitimate democracy. A limited number of people should be interested enough in an issue to build up the necessary knowledge and devote the time to promote the cause. If social media and other digital media make it easier for motivated people to connect to each other, this is probably a good thing for democracy. But further empirical studies must take into account the old question of whether the elites, new or old, have views that are representative of the people as a whole.

If politics is the art of the possible, we, as social scientists, should not mourn the seeming impossibility of mass activism spread equally through all fractions of society, but critically assess new forms of political organization in their societal context. We must compare emergent developments in democracy with reality, not with abstract democratic ideals.

Notes

1. "Leetocracy" means basically "rule of the leet." "Leet" is an Anglo-American internet slang term deriving from the word "elite." It denotes the special kind of language used by hacking and other online cultures, using abbreviations (lol, brb), numbers instead of letters (1337) and deliberate misspelling (pwned), as well as the self-appointed digital elite using the language. We use the term "leetocracy" as a reminder that the increased importance of networked political activism might not necessarily mean increased equality in political participation but instead the potential rise of new elites as argued in this paper.
2. La Libre Belgique, 2009, http://bourse.lalibre.be/actualites.html?id=20090308T120621 Z&genre=AFP&ticker=&pays=&source=afp (last accessed 10/01/2010)
3. http://www.laquadrature.net/en/faq-0 (last accessed 10/01/2010)
4. This paper draws on a series of 20 interviews conducted with activists and members or staff of the European Parliament involved in the Telecoms Package reform. All interviews and data collection were carried out between September 2008 and December 2009, and the sources have been analyzed following a thematic, inductive inspection. Documents generated by the activists themselves such as press releases, analyses and further documentation posted on their website and wiki, messages posted on mailing lists and documents, and analyses provided by the activists themselves or the political staff inside the EP were also taken into account.
5. http://www.laquadrature.net/en/who-are-we (last accessed 10/01/2010)
6. Ibid.
7. "A patch is a small piece of software designed to fix problems with or update a computer program or its supporting data." (Wikipedia "Patch," http://en.wikipedia.org/wiki/Patch_(computing), last accessed 10/01/2010)
8. La Quadrature du Net launched an internet blackout, that is, the voluntary dressing in black of websites, avatars, etc., in order to influence the French legislative proposal, the HADOPI law (see below).
9. The HADOPI law is the acronym used for the Loi n°2009–669 du 12 juin 2009 favorisant la diffusion et la protection de la création sur internet (Law n°2006–660 of June 12, 2009 facilitating the diffusion and protection of creation on the internet) implementing the three strikes mechanism in France.
10. http://www.laquadrature.net/lawtracks/telecoms_package/ (last accessed 10/01/2010)
11. http://eur-lex.europa.eu/en/index.htm (last accessed 10/01/2010)

References

Barber, B. (1984). *Strong democracy: Participatory politics for a new age.* California, US: University of California Press.
Baringhorst, S. (2009). "Introduction: Political campaigning in changing media cultures—Typological and historical approaches." In Baringhorst, S., Kneip, V., Niesyto, J (eds.), *Political campaigning on the web.* (pp. 9–30). Bielefeld, Germany: Transcript Verlag.
Beck, U. (1997). *The reinvention of politics.* Cambridge, UK: Polity.
Becker, T.L., & Slaton, C.D. (2000). *The future of teledemocracy.* Westport, CT: Praeger
Beitz, C. R. (1989). *Political equality.* Princeton, NJ: Princeton University Press.
Benkler, Y. (2006). *The wealth of networks.* New Haven, CT: Yale University Press.
Bennett, W.L. (2004). Communicating global activism: Strengths and vulnerabilities of networked politics. In Van De Donk, W., Loader, B.D., Nixon, P.G. & Rucht, D. (Eds.) (2004). *Cyberprotest: New media, citizens, and social movements* (pp. 123–146). London: Routledge.
Bentivegna, S. (2006). Rethinking politics in the worlds of ICT. *European Journal of Communication, 21*(3), 331–343.
Berry, D. M. (2008). *Copy, rip, burn: The politics of copyleft and open source.* London: Pluto.
Breindl, Y., & Briatte, F., (2009). Activisme sur internet et discours stratégiques autour de la propriété intellectuelle. *Terminal, technologie de l'information, culture et societé: Un nouvel activisme sur Internet,* n° 103–104, 23–32.
Breindl, Y., (2010). Internet-based protest in European policy-making: The case of digital activism. *International Journal of E-politics,* forthcoming.
Castells, M. (1996). *The Rise of the Network Society.* Cambridge, MA: Blackwell.
Castells, M. (2001). *The internet galaxy.* Oxford, UK: Oxford University Press.
Chadwick, A. (2006). *Internet politics: States, citizens, and new communication technologies.* New York, NY: Oxford University Press.
Corbineau, B., & Barchechath, E. (2003). The discourse on e-democracy: Where are we heading? In *Building the knowledge economy: Issues, applications, case studies.* Oxford, UK: IOS.
Dahlgren, P. (2004) 'Foreword.' In W. van Donk, B.D. Loader, P.G. Nixon, & D. Rucht (Eds.), *Cyberprotest: New media, citizens, and social movements* (pp. 11–16). London and New York: Routledge.
Dahlgren, P. (2009). *Media and political engagement: Citizens, communication, and democracy.* Cambridge, US: Cambridge University Press.
Dalton, R. J., (2008). 'Citizenship norms and the expansion of political participation.' *Political Studies, 56,* 76–98.
Downs, A. (1957). *An economic theory of democracy.* New York: Harper Collins.
Earl, J., & Schussman, A. (2003). The new site of activism: Online organizations, movement entrepreneurs, and the changing location of social movement decision making. *Consensus Decision Making: Northern Ireland and Indigenous Movements, 24,* 155–187.
Fishkin, J. (1991). *Democracy and deliberation.* New Haven, CT: Yale University Press.
Flichy, P. (2001). *L'imaginaire d'Internet,* Paris: La Découverte.
Giddens, A. (1991). *The consequence of modernity.* Cambridge, UK: Polity.
Gustafsson, N. (2009). 'This time it's personal: Social networks, viral politics and identity management.' In Riha, D., Maj, A. (Eds.), *The real and the virtual.* Oxford, UK: Inter-Disciplinary Press.
Habermas, J. (1989). *The structural transformation of the public sphere.* Cambridge, MA: The MIT Press.

Habermas, J. (1996). *Between facts and norms.* Cambridge, MA: The MIT Press.
Held, D. (2006). *Models of democracy.* Oxford, UK: Blackwell.
Hindman, M., (2008). *The myth of digital democracy.* Princeton, NJ: Princeton University Press.
Hoff, J., & Bjerke, F. (2009). *How should we understand the relationship between internet and politics? Towards a theoretical framework,* Paper presented at the ECPR General Conference, Potsdam Germany, 10–12 September 2009.
Inglehart, R., (1977). *The silent revolution: Changing values and political styles among Western publics.* Princeton, NJ: Princeton University Press.
Jenkins, H., (2006). *Convergence: Where old and new media collide.* New York: New York University Press.
Jensen, J. L., (2006). The Minnesota e-democracy project; Mobilizing the mobilized? In *Internet and Politics* (pp. 39–58). Oxon, UK: Routledge.
Joyce, M., (2007). 'Civic engagement and the internet: Online volunteers.' *Internet and democracy blog,* 2007-11-18. Retrieved 3rd January, 2009, fromhttp://blogs.law.harvard.edu/idblog/2007/11/18/civic-engagement-and-the-internet-online-volunteers/
Lévy, P. (1997). *Collective intelligence: Mankind's emerging world in cyberspace.* Cambridge, MA: Perseus.
Lévy, P. (2002). *Cyberdémocratie: essai de philosophie politique.* Paris, France: Odile Jacob.
Levy, S. (1984). *Hackers: Heroes of the computer revolution,* London: Penguin.
Micheletti. M., (2003). *Political virtue and shopping: Individuals, consumerism, and collective action.* Basingstoke, UK: Palgrave Macmillan.
Miller, D., (1983). 'The competitive model of democracy.' In Duncan, G. (Ed.), *Democratic theory and practice.* Cambridge, UK: Cambridge University Press.
Morris, D., & Delafon, G. (2002). *Vote.com ou comment internet va révolutionner la politique.* Plon, Mesnil-sur- l'Estrée, France.
Nonnecke, B. and Preece, J. (2000). Lurker Demographics: Counting the Silent. *Proceedings of CHI'2000,* Den Haag, The Netherlands, 73–80.
Norris, P., (2002). *Democratic phoenix. Reinventing political activism.* Cambridge, MA: Cambridge University Press.
Pateman, C., (1970). *Participation and democratic theory.* Cambridge, MA: Cambridge University Press.
Putnam, R., (1976). *The comparative study of political elites.* Englewood Cliffs, NJ: Prentice Hall.
Putnam, R., (2000). *Bowling alone: The collapse and revival of American community.* New York: Simon & Schuster.
Rasmussen, T. (2007). *Techno-politics, internet governance and some challenges facing the internet,* Oxford Internet Institute, Research Report 15. Retrieved 4th November, 2009 from http://ssrn.com/abstract=1326428
Rebentisch, A., (2005). Eine Fuehrung in das europaeische IT-lobbying, 22nd Chaos Computer Congres, Berlin.
Rheingold, H. (2000). *The virtual community: Homesteading on the electronic frontier* (2nd ed.). Cambridge, MA: The MIT Press.
Rheingold, H. (2002). *Smart mobs: The next social revolution.* Cambridge, MA: Perseus.
Schumpeter, J. (1946). *Capitalism, socialism, and democracy.* New York: Harper & Row.
Shirky, C., (2008). *Here comes everybody. The power of organizing without organizations.* New York,: Penguin.
Van de Donk, W. B. H. J., Snellen, I. Th. M., & Tops, P. W. (Eds.) (1995). *Orwell in Athens. A perspective on informatization and democracy.* Amsterdam: IOS.
Vanobberghen, W. (2007). 'The marvel of our time': visions surrounding the introduction of radio broadcasting in Belgium 1923–1928, Paper presented at Media History and History in the Media: Media and Time, Wales, Gregynog (University of Aberyswyth), 28–30 March 2007.

Verba, S., Schlozman, K., & Brady, H. (1995). *Voice and equality. Civic voluntarism in American politics.* Harvard, MA: Harvard University Press.

Wellman, B., Haase, A. Q., Witte, J., & Hampton, K., (2001). 'Does the internet increase, decrease or supplement social capital? Social networks, participation, and community committment,' *American Behavioral Scientist. 45*(3), 437–456.

Yu, Peter K., (2004). The escalating copyright wars. *Hofstra Law Review, 32*, 907–951.

CHAPTER TEN

Haxorz[1]

Alternate Perspectives on the "Computer Underground"

TESSA J. HOUGHTON & YAO-CHUNG CHANG

INTRODUCTION

"Hacking is undoubtedly one of the buzzwords of the computer age" (Vegh, 2003, p. 151). Yet the definition of the action, the hacktion, and its actor, the hacker, remains mercurial. The origin of the terms and the identity they refer to is obscure, as they have been and are used in a variety of contexts and connotations. Despite ongoing disagreement over even its most basic level of meaning, hacking is arguably best understood along the lines of a playful, creative, ingenious interaction, modification, or manipulation of an object or subject. A certain spirit inherent in the activity remains constant, but the actor and that which is acted upon do not. Fashion, craft, architecture—any aspect of life can, theoretically, be hacked. People or "wetware" can be hacked, hardware can be hacked, and, of course, software and networks can also be hacked.

These latter types of "hackable objects" (electronic and specifically computing hardware, software, and networks) support the most common general meaning of "hacking," but even this narrowed categorisation contains fractures and contradictions. As described in the following perspectives, "computer hacking" has been and is used to mean rather different things. Are hackers "heroes of the computer revolution" (Levy, 1984), or are they "electronic bogeymen"? Do they hack for fun, "kudos," political reasons, or out of greed and malignancy? Is what they do illegal, or does it represent a particular work ethic and means of production? The answer is, arguably, all of the above, and more. Computer hacking has been occurring for six decades, and the practice has evolved in numerous directions, for numerous reasons, and with numerous outcomes. It is highly heterogeneous and polysemic.

This floating polysemy leaves the terms hacking and hacker vulnerable to attempts at external meaning making and fixing. Governments and corporate media often frame hackers as purely criminals or terrorists, conflating the term with the practices of "cracking" and cyberterrorism. These illegal and destructive practices are arguably best understood as particular branches on the evolutionary tree of computer hacking, but they are not representative of the practice at large. The programmers "hacking up" code for money or love, for proprietary or free/open source projects, at work, at home, and in hackspaces, and the "ethical hackers" or computer security professionals, are all hackers. They almost certainly comprise the vast bulk of the global hacker community in terms of sheer numbers. They are simply not as "newsworthy" as those hacking into a power plant's SCADA system or a bank's credit card records, or into Google, Adobe, or Intel's corporate network.

The following two perspectives explore two of the evolutionary extremes of hacking: hacktivism and cybercrime (particularly cracking for financial gain). Hacktivists, particularly the "political crackers" Houghton (perspective 1) focuses on, may use techniques very close to those utilised by financially motivated cybercriminals, and their activity is similarly illegal. However, the cybercriminals discussed by Chang (perspective 2) are motivated largely by profit, whereas hacktivists hack in aid of political causes. Houghton approaches the topic through the lens of democratic theory, asserting that hacktivists should be understood as legitimate contributors to a global network of public and counterpublic spheres, in that they transform violent antagonism into expressive agonism, thus diffusing the conflict inherent to "the political" (Mouffe, 2005). Chang, coming from a legal and criminological background, explores the difficulties involved in legislating effectively against cybercrime, in light of the transnational nature of hacking, the national nature of legislation, and the difficulty of actually tracking and apprehending cybercriminals.

These contrasting perspectives are intended to highlight the diversity of the practice of hacking, and to prompt a more holistic understanding of it—one that extends beyond the sensationalist electronic criminal stereotype, without overcorrecting towards a rose-tinted counterpart. Hacking should be understood as a practice, approach, or toolkit. What it is ultimately used *for* can vary widely. It is a means, not necessarily an end in and of itself (although most hackers do see hacking as an inherently enjoyable). In a world where postindustrial societies are increasingly networked and online, the hacker assumes a heightened status and importance, as one who can expertly navigate this computerised domain. Rather than fearing and demonising this expertise, we should rather attempt to understand it in all its permutations, learning from it where and when we can.

PERSPECTIVE 1:

The People's Republic of Hacktivism: A Public Sphere Theoretical Interpretation of Online Independence Movements and the People's Republic of China
Tessa J. Houghton

The People's Republic of China is widely recognised as one of the world's hacking "hotspots." This activity is often discussed in terms of cyberespionage, cyberterrorism, or cyberwarfare, but there is less recognition of hacktivism in the region. This chapter defines hacktivism as a phenomenon by assessing its history and typology, and differentiating it from cyberwarfare and cyberterrorism. Two case studies involving hacktivist events between the People's Republic of China and regional independence movements—Taiwanese and Uighur—are used to provide a template for understanding hacktivism and situating it as a legitimate form of public sphere activity. A neo-Habermasian conception of the public sphere is defined and applied to both case studies to elucidate the politically legitimate role hacktivism plays in a multi-tiered global "modular network" of counterpublics (Keane, 2000, p. 87). As such, this chapter provides a template for the application of neo-Habermasian theory to a wider range of creative forms of political expression and communication.

THE NEO-HABERMASIAN PUBLIC SPHERE

Public sphere theory is embedded within the wider context of deliberative democratic theory, which prioritises the transformation of individual preferences through heterogenous deliberation, as opposed to their mere aggregation. This process requires participants to interrogate and justify their preferences, thus raising the quality of decision-making. This shift in focus reorients our perspective on political "work." The focus is less on national governmental activity, and more on the communicative work done by and within civil society. The increasing globalisation of civil society and political power structures (facilitated by the development of global communication systems) makes it "far from obvious that the myopic focus on state power is any longer justified—if it ever was" (Shapiro, 2004, p. 13). Participatory capital has relocated into new concerns and modes, rooted more firmly in global civil society than in governmental institutions, and encompassing expressive as well as instrumental activity (Dahlgren, 2007).

Public sphere theory generates ideal conditions for and boundaries to the kinds of deliberation regarded as politically legitimate. It is sometimes treated as a stabilised discourse, but the Habermasian concept of the public sphere, both in its original form (1989) and later versions, has provoked an immense amount of criticism and reformulation. The public sphere's social embeddedness necessitates constant reformulation—it must keep pace with society (Dahlgren, 1991). A "neo-Habermasian" understanding of the public sphere is proposed, responding to Habermas's concept and associated criticisms, and building upon the "agonistic," or "radical" theoretical tradition. It expands and sensitises the concept, moving away from rational-critical preoccupations, thus comprehending issues of power and difference, and allowing publicness "to navigate through wider and wilder territory" (Ryan, 1992, p. 286).

Neo-Habermasian public sphere theory centres on the notion of multiple counterpublic spheres, *à la* Nancy Fraser (1992). These counterpublics define themselves through the discursive struggles they elect to take part in, with discourses understood as "shared set[s] of concepts, categories, and ideas that provide [their] adherents with a framework for making sense of situations, embodying judgements, assumptions, capabilities, dispositions, and intentions" (Dryzek, 2006, p. 1). They are defined oppositionally: against one another, or against dominant or hegemonic spheres as counter-hegemonic projects (Dahlberg, 2007). Dominant spheres, although nymically defined as public, may in fact be pseudo-publics of the kind explored in Habermas's discussion of the decline of the bourgeois public sphere and the wide-ranging refeudalisation of society (1989).

Publics range from subnational to supranational, as in Keane's theory of micro, macro-, and meso-public spheres, which visualizes "a complex mosaic of differently-sized, overlapping and interconnected public spheres" (2000, p. 76). As Keane argues, "[t]he ideal of a unified public sphere and its corresponding vision of a territorially bounded republic of citizens striving to live up to their definition of the public good are obsolete" (ibid.). We must rather imagine a multi-tiered global "modular network" (ibid., p. 87) of counterpublics, linked by flows of resistance and opposition, as well as "chains of equivalence" (Laclau & Mouffe, 1985; Mouffe, 2005) of varying strengths and types. These "chains" are connections based on elements of resonance and similarity between different discourses. They form a collective "we" that resists being "neutralised in the marketplace of multicultural pluralism, or polarised in a reductive competition of victimisations" (Downey and Fenton, 2003, p. 194). This transnationalisation is necessary, as "the public sphere has been reshaped through the globalising, mediated forms of communication that constitute the representational infrastructure for today's public spaces" (McLaughlin, 2004, p. 157).

Publics have both outward and inward, group-solidarity based orientations. External "publicity" need not be oriented towards governments or seeking state power (Dryzek, 2000), or even any kind of formal institution. It may be aimed at non-institutional power structures and elites, and seeking influence within civil society. Counterpublic discourses are politically legitimised through their very articulation, not prescribed *a priori* by governing elites (Ryan, 1992). The legitimisation of identity and "difference" centred concerns has expanded the concept of the "political" (Benhabib, 1996), and this openness to new issues and allowing for the "politics of the personal" to become "public" is essential to maintaining the public sphere's core ideal of inclusivity.

Defining which modes of deliberation are politically legitimate is fundamentally important to public sphere theory. Habermas and many other theorists hold to the ideal of rational-critical procedural forms of judgement, which have been roundly criticised. The notion of "status-bracketing" as a means of ensuring deliberative equality is fundamentally exclusionary—the ability to operate as a rational-critical equal is a differential resource. Legitimising the deliberation of only those possessing this privileged ability serves to simultaneously obscure and reify existing power relations, rather than facilitate any critical political function (Dahlberg, 2007; Mouffe, 2005; Ryan, 1992). As such, neo-Habermasian public sphere theory discards rational-criticality as counterproductive. Various modes of deliberation are still judged on how well they fulfil deliberative authenticity, but they are instead required to induce self-reflection on political preferences in a non-coercive fashion, as proposed by Dryzek (2000).

Furthermore, the achievement of truly rational consensus is regarded as ontologically impossible—it eliminates plurality, a fundamental component of the public sphere, and is based upon hegemonic discursive stabilisation (Mouffe, 2005). As Mouffe argues, "[p]luralism is not merely a *fact*, something that we must bear grudgingly or try to reduce, but an axiological principle…[and] something we should celebrate and enhance" (2000b, p. 19). Power is an ineradicable component of social relations, because all social objectivities are premised upon constitutive exclusion—every "we" bears traces of the "they" it is defined against (Mouffe, 1993).

We should instead seek to defuse and transform violent antagonism between enemies into agonism between adversaries—"somebody whose ideas we combat but whose right to defend those ideas we do not put into question" (Mouffe, 2000a, p. 15). We acknowledge adversaries as legitimate opponents, but are unable to come to an agreement. A certain amount of consensus based around a dedication to the values of liberty and equality is necessary. However, it can only ever be a "conflict-ual consensus" given that the interpretations of these values will be so varied and irreconcilable, and should always remain open to contestation. The process of deliberation and contestation is infinitely more valuable than any of the temporary and flawed "consensual conclusions" to this process (Dahlberg, 2005).

As long as activism induces preference reflection in a non-coercive fashion, and transforms antagonism into agonism, it is a legitimate component of the neo-Habermasian global modular network of counterpublics. It makes us "wonder about what we are doing, [and] rupture[s] a stream of thought, rather than [weaving] an argument" (Young, 2001, p. 687). This is true for activism carried out both off- and online, therefore the "repertoires of electronic contention" utilizing "conventional" and "disruptive" tactics categorised by Costanza-Chock (2001) constitute different modes of internet-based counterpublic spheres. This latter "disruptive" category encompasses various tactics or "cultural outcomes," all of which are collectively known as hacktivism.

Hacktivism fulfils the neo-Habermasian requirement of provoking non-coercive reflection on political preferences, and transforming antagonism into agonism. As Dahlberg contends, "protest is very much a communicative act when undertaken with the aim of raising issues for deliberation rather than to coerce." It utilises "creative and sometimes 'disruptive' forms of rhetoric through which marginalised groups can gain a hearing for their voices and call into question more dominant positions." (2005, p. 119–120).

HACKTIVISM: A GENEALOGY AND TYPOLOGY

Hacktivism is an amalgamation of the techniques and ideologies of traditional activism or protest with the techniques and ideologies of hacking. The combination may seem odd at first glance, due to the mass-mediated pathologisation of hacking (Nissenbaum, 2004). In the public consciousness, the term "hacker" has come to signify an "electronic bogeyman"—a malicious intruder who breaks into computer networks to wreak havoc and destruction for personal gain (be it money or infamy). While this description may fit the subset of hacking known as "cracking," hacking itself is a much more diverse and ethical tradition than this hackneyed stereotype expresses.

The Evolution of Hacking

Levy (1984) provides a history of the early years of hacking. He defines the first three "generations" of hackers, from the early computing pioneers at major US educational institutions, through to the hardware hackers who contributed to bringing personal computing to the general public and the software or game hackers who followed them. He also defines a list of "hacker ethics" that have enduring relevance as an articulation of "the complex construction of a collective identity" (Jordan & Taylor, 1998, p. 775). They state that access to computers should be unlimited and total, all information should be free, authority should be mistrusted and decentralisation promoted, hackers should be judged by their hacking alone, art and beauty can be created on computers, and that computers can change your life for the better (Levy, 1984, p. 27–33). Although Turkle's (1984) concurrent description of "the hack" identified illicitness—the idea that the hack "must be against some institutional, legal, or even just perceived rules" (p. 236),—as an integral component to hacking, this component need not be malicious, destructive, or lead to financial gain.

Jordan and Taylor (2004) continue this genealogy[2], identifying the "hacker/cracker" conflation beginning in the eighties with the emergence and commodification of cracking activity, and the rise of the "Microserf"—hackers and programmers co-opted by corporations such as Microsoft. This era constituted the "nadir of the original hacker ethic" (Taylor, 2001, p. 63), but was followed by the most explicitly political generations—the free/libre and open source software (FLOSS) movement and hacktivism. These generations represent a retreat from commodification and the "concomitant reassertion of more countercultural values" (Jordan & Taylor, 2004, p. 15–16).

Hacktivism

This history explains the dovetail between hacking and socio-political activism. Hacking has long had its own implicit political ideology which, when coupled with the increasing digitisation of politics, economies, and society, demystifies the emergence of hacktivism. Hacktivism must be understood not only for what it is, but also for what it is not—that is, it is not cyberterrorism or cyberwarfare, although mainstream media coverage displays little knowledge of these distinctions. 9/11 brought an increasing media and legislative conflation of these activities (Vegh, 2003). However, their perpetrators and effects are quite different.

Hacktivism is demonstrably ethically motivated, and is disruptive without resulting in human, infrastructural, or serious financial casualties (unlike cyberwarfare or cyberterrorism). It may be peripherally aligned with an ongoing conventional armed conflict, but its perpetrators have no direct governmental, military, or diplomatic connections. In contrast, cyberwarfare is directly connected to ongoing conventional armed conflicts and is perpetrated by the governments and militaries involved in them. An example is the 2008 crippling of the Georgian communication system, timed to coincide with the Russian military strike and apparently committed by government-directed hackers, as revealed in the "Grey Goose" reports

("Grey Goose 2," 2009). Finally, the definition and indeed existence of cyberterrorism is somewhat contested ("Cyberterrorism," 2009), but it is essentially "the convergence of cyberspace and terrorism. It covers politically motivated hacking operations intended to cause grave harm such as loss of life or severe economic damage" (Denning, 2001, p. 241).

Different Trends within Hacktivism

Hacktivism also has an internal typology. The convergence of increasingly politically aware hackers and increasingly technologically savvy activists has led to distinct trends—"mass action" (MAH) and "digitally correct" (DCH) hacktivism (Jordan &Taylor, 2004). MAH draws upon street protest and civil disobedience and *defies* "the lack of physicality in online life, in favour of a mass collection of virtual bodies that are not yet present to one another" (ibid., p. 69). Virtual sit-ins and satirical performance-based hacktions such as the parody mirror sites of ®™ark[3] and the Yes Men,[4] are MAH. DCH, in contrast, *uses* "the lack of physicality in online life to amplify a political message" (ibid., p. 6). True to its hacker roots, it is often in aid of "bandwidth rights," which sometimes brings it into conflict with MAH. The group "Hacktivismo,"[5] with its overarching aim of circumventing internet censorship, exemplifies DCH.

Orientation/Origin	Hacker-Programmer Origin	Artist-Activist Origin
Transgressive Orientation	'Political Coding' *(DCH)*	'Performative Hacktivism' *(MAH)*
Outlaw Orientation	'Political Cracking'	

Figure 1. Samuels' (2004) typology of hacktivism, with Jordan & Taylor's (2004) categories inserted. The case studies are located in the shaded portion of the matrix.

As apparent in Figure 1, the gap in Jordan and Taylor's argument—accounting for defacements and individually undertaken hacktions—is closed by Samuels' taxonomy. She creates a matrix of hacktivist origins (hacker-programmer or artist-activist) and orientations (transgressive—challenging the legal and political order but still existing in relation to it, or outlaw—completely rejecting the legal and political order) (2004, p. 37). "Performative hacktivism" (MAH) is done largely by artist-activists, and takes transgressive forms such as site parodies and virtual sit-ins. "Political coding" (DCH) is done largely by hacker-programmers involved in transgressive software development. Her additional category, "political cracking," is usually carried out by hacker-programmers, and consists of outlaw permutations such as DDoS attacks, information theft, the use of viruses/trojans, and site alterations/redirections. Although the usage of the term "cracking" is arguably inappropriate, the categories themselves provide a sophisticated typology. They are not rigid—the potential for overlap is recognised (2004, p. 3–4)—but they provide an invaluable interpretive guide.

TERRITORIAL POLITICAL CRACKING: TAIWAN, THE UIGHUR, AND THE PEOPLE'S REPUBLIC OF CHINA[6]

Hacktivism is a diverse activity, with many possible permutations and driving discourses, and examining it in general is impossible here. However, an introduction to establishing it as a legitimate form of neo-Habermasian public sphere activity is possible. This perspective prompts a more general acceptance and exploration of various forms of creative communication as legitimate contributions to political discourse, and counters continued allegiances to the rational-critical procedural model. An analysis of a particular category of hacktion (political cracking) and discursive struggle (the ongoing territorial conflicts centred around the People's Republic of China) is developed through two case studies. The choice of political cracking is purposive—due to its "outlaw" orientation, it is the category of hacktivism that appears most at odds with public sphere theory.

The first case is that of the July 1999 site defacements following the Taiwanese President's description of Taiwan's relations with mainland China as a "special state-to-state relationship"; and the second, the recent defacement of Australian and Taiwanese film festival websites over plans to screen *10 Conditions of Love*, a documentary about exiled Uighur activist Rebiya Kadeer. These two incidents are part of a larger conflict centred on territorial issues in the People's Republic of China (PRC), and in particular, the specific conflicts between mainland China and Taiwan over whether Taiwan is an independent state; and in the Xinjiang Uighur Autonomous Region (XUAR) over the Uighur separatist or independence movement.

The PRC and Taiwan: "One China" versus Taiwanese Independence

On July 9, 1999, Taiwanese President Lee Teng-hui stated in an interview with *Deutsch Welle* that since the introduction of its constitutional reform in 1991, Taiwan had redefined its interactions with mainland China as being "state-to-state, or at least special state-to-state relations" ("President Lee Teng-hui's responses," 1999). This statement led to immediate condemnation from Beijing. An official response stated that "[t]here is only one China in the world and Taiwan is part of Chinese territory," and that Lee's "separatist malice" was "defying the will of the people" ("Spokesman," 1999).

Lee's statement and the PRC's disapproving reaffirmation of "One China" policy also drew the attention of Chinese hackers, who defaced several Taiwanese governmental websites, including those of the National Assembly and the Control and Administrative Yuan (thus defacing three of what were then the six branches of the Taiwanese government). Usual content was replaced with images of the Chinese flag, and political slogans stating that Taiwan was "an undividable part of China." However, their Taiwanese counterparts quickly retorted in kind, defacing PRC governmental websites with pro-Taiwanese independence messages and adding insult to injury with the addition of "liberal" pop-cultural references, including pop songs and animations ("China, Taiwan," 1999).

Given the ongoing conflict between mainland China and Taiwan and the inter-governmental dialogue in this situation, it is tempting to suspect both groups of hackers of operating under governmental directives, but there is little evidence to support this. When Taiwanese authorities warned that the anti-PRC defacements were illegal and might result in prosecution, the Taiwanese hackers responded that they were an autonomous movement against "information warfare." Despite the "popular media conception [being] that there is a coordinated attempt by the Chinese government to hack into U.S. computers," US computer security expert Bruce Schneier believes that the vast number of Chinese hackers hacking pro-Tibet, pro-Taiwan, Falun Gong, and pro-Uighur as well as US sites, are generally not working for the PRC government. "They're basically young, male, patriotic Chinese citizens, trying to demonstrate that they're just as good as everyone else ... [and] are in this for two reasons: fame and glory, and an attempt to make a living. They're upholding the country's honor against both anti-Chinese forces like the pro-Tibet movement and larger forces like the United States" (2008).

The lack of evidence of government involvement and of any simultaneous armed conflict between China and Taiwan, and the fact that the defacements were politically oriented but caused no grave harm to online or offline infrastructures (and certainly never threatened any human lives), classes this incident as hacktivism rather than cyberwarfare or cyberterrorism. It is an example of interactions between neo-Habermasian public spheres, taking part in a wider discursive struggle over the geopolitical status of Taiwan. The defacements are intended to provoke their target into reflecting upon political preferences in a non-coercive fashion—they are "'disruptive' forms of rhetoric through which marginalised groups can gain a hearing for their voices and call into question more dominant positions" (Dahlberg, 2005, pp. 119–120).

There are two possible interpretations of this incident with regards to defining the approximate discursive boundaries and dominant/counter-status of the publics involved. The first would be to perceive the "One China" discourse as the dominant regional public, and Lee Teng-hui's assertion of independence as the first Taiwanese counterpublicity against this dominance. The assertion threatened "One China's" discursive stability, and the ensuing hacktivism "hacked into" the Taiwanese counterpublic, in order to destabilise it in turn and reassert the "One China" discourse. The reactive Taiwanese counterpublic hacktivism "hacked into" the "One China" discursive public in a similar fashion, seeking to simultaneously destabilise its adversary and restabilise itself. The other interpretation is to view both discourses or spheres as dominant within their own nation states. Both sets of hacktivism are then the actions of marginalised counterpublics "hacking into" the respective dominant national publics to destabilise their discourses through provoking political preference reflection among their members.

Either interpretation is valid—the critical issue is that the hacktions between China and Taiwan were legitimate deliberative events within an interplay between two public spheres, with both spheres and discourses regarding each other as mutually oppositional or "counter." Although China is internally non-democratic, this counterpublic interaction was a specific temporal and political node within the "modular network" of global deliberative democracy envisaged by neo-Habermasian public sphere theory. The discourses of "One China" and Taiwanese independence define themselves in relation to one another, and the power structures and exclusion inherent in them are unable to be reconciled. As Mouffe argues, this pluralism "is not merely a *fact*, something that we must bear grudgingly or try to reduce, but an axiological principle . . . of modern democracy" (2000b, p. 19). The mutual antagonism between the discourses cannot be eliminated, only defused and transformed into agonism.

The defacements achieved this conversion through the provocation of political preference reflection in members of both discourses or publics in a non-coercive manner—they were temporarily disruptive but did not violently force preference alteration, and did not seek to annihilate enemies, but to dispute adversarial discourses. Both PRC and Taiwanese citizens felt that their discourses (nationalism and independence) were not being adequately heard by their opposing counterpublics. Hacking into each other's national discourses surmounted this perceived lack of engagement and inter-public activity, and ensured that their voices were heard outside their own public spheres. This interplay is characteristic of hacktivism as a counterpublic activity, and is one that was replicated more recently with regards to the showing of the film *The Ten Conditions of Love*.

The PRC and the XUIR: The 10 Conditions of Love

The Ten Conditions of Love (*TCoL*), a documentary about exiled Uighur activist, Rebiya Kadeer, was scheduled as part of the Kaohsiung Film Festival, Taiwan, and the Melbourne International Film Festival, Australia. *TCoL* follows Kadeer's family, and their efforts to gain more autonomy for the approximately 10 million mainly Muslim Uighurs living in the Xinjiang Uighur Autonomous Region (XUAR) of the PRC. The screenings came only months after ethnic violence in the region resulted in the deaths of about 200 Muslim Uighurs and Han Chinese, and relations in the region remain tense. Chinese authorities have accused Kadeer and the World Uighur Congress (which she heads) of inciting the violence, a charge Kadeer has repeatedly denied. The inclusion of *TCoL* in both festivals' line-ups led to various forms of condemnation from the PRC, including nationalist hacktivism.

The Melbourne festival was hit in July, after the Chinese government demanded that the film be pulled. After festival organisers refused, several Chinese directors pulled their films from the festival, with one stating, "I do not want to see my film screened on the same platform as a film about Kadeer" (Tran, 2009). This censure was followed by a series of defacements to the festival site, replacing programme information with the Chinese flag and anti-Kadeer slogans (see Figure 1), as well as an email flood and a ticketing form flood. The Kaohsiung festival met with similar problems in September—the site was defaced with a digitally altered image of Kadeer and the Dalai Lama, in reference to frequent comparisons between the two as exiled leaders and the Dalai Lama's recent visit to Taiwan. Pro-PRC slogans were again present, with one declaring "anti-Xinjiang, anti-Tibet and anti-Taiwan independence. Fervently celebrate the motherland's 60th anniversary!" (Child, 2009), and a message on an associated blog reiterated Kadeer's alleged culpability for the bloodshed in the XUAR.

Again, there is no evidence that the actions were from governmental quarters, and likely came from patriotic PRC citizen-hackers. "Oldjun" is a well-known Chinese hacker and hacktivist, and was one of those involved in both the Melbourne (see Figure 2) and Taiwanese festival defacements. In an interview with the China Daily Monday, he stated explicitly that "I hacked into [both] their website[s] because I simply want to express people's anger about the screening of the film....It is my own doing. Nobody told me to do it. I really don't understand why they have to show the film." (Jia, 2009).

The defacements and floods were disruptive, but did not violently force preference reflection or seek to annihilate enemies, but to dispute adversarial discourses. The hacktivism was again part of an attempt to restabilise a wider Chinese nationalist public sphere after it had been destabilised by counterpublic activity. In this case, the primary counterpublic was constituted by the Uighur independence discourse articulated through *TCoL* and its associated publicity, but there was also an extra element in play. The screening of the film by Australia and Taiwan, and their resistance to official Chinese censure, represents the articulation of a larger networked counterpublic, with "chains of equivalence" emerging between the Uighur, the Australian film community and Taiwan. In this case, an implicit chain was built between the counterpublics or discourses of the Uighur independence movement, the Taiwanese independence movement, and with a global frame defending free speech and human rights. This served to amplify the different voices constituting it, ensuring that their combined influence and solidarity was greater than the sum of its parts. The PRC defacement slogan decrying the Uighur and Taiwanese independence movements clearly recognises these connections, and perceives the Tibetan movement to be a further component (likely because of the Dalai Lama's visit to Taiwan), seeing them as a combined threat to the PRC nationalist discourse.

Figure 2. One of the Melbourne International Film Festival defacements[7]

The second point of note is that the Chinese hacktivism certainly provoked self-reflection on political preferences, but it would seem that this reflection was closely followed by stalwart reassertion within the counterpublics it "hacked into." Despite the widespread disruptions to the Melbourne festival, its directors used the accompanying publicity to reassert their autonomy and belief in the importance of the film, and ticket sales were strong, with the directors even considering adding more sessions at one point. In Kaohsiung, *TCoL* was pulled from the festival, but was shown early at the governmental Kaohsiung Film Archive, to ensure the festival was not further affected without compromising dissemination of the film. A Taiwanese editorial outlined the importance of the decision: in showing the film, Taiwan would demonstrate its commitment to democracy and freedom of expression, and criticise the PRC's "brutal handling of demonstrations for autonomy in Tibet and Xinjiang or other violations of human rights" ("Why Taiwan must show *TCoL*," 2009).

Discussion and Concluding Remarks

Evidently, despite hacktivism being a legitimate form of neo-Habermasian public sphere activity, it is not guaranteed to be successful in converting or transforming political preferences. The use of political cracking, with its outlaw orientation, may alienate its perpetrators and their discourses. When used against smaller targets such as the Uighur and Taiwan by a public such as the PRC, who are generally perceived as powerful or dominant, there is a strong likelihood that it will be perceived as "bullying" by both the counterpublic targets and the wider global "public of publics." This perception is likely to undermine any chances that the provocation of political preference reflection will result in the transformation of said preferences, and may in fact cause the counterpublics to retrench their beliefs and expressions. Furthermore, the consequent defensive and victimised framing of these counterpublics may lead to a more global condemnation of the discourse of the regionally dominant public, especially when this public is internally undemocratic or is believed to be less than progressive in various other respects. Essentially, the inwards, group-solidarity oriented component of the dominant public may be internally successful (and indeed, this may at times be the primary goal), but its externally oriented publicity campaign may "fail."

This discussion of hacktivism in general as a legitimate form of neo-Habermasian public sphere activity has been but a brief introduction. Nevertheless, the case studies demonstrate the value of interpreting hacktivism through a neo-Habermasian lens. Due to its "outlaw" orientation, political cracking, in these cases taking the form of defacements and floods, is the form of hacktivism that pushes the boundaries most strongly in terms of whether or not it induces political preference reflection in a non-coercive manner. As such, any discussion of it serves to test the application of the theory. The analyses of the 1999 conflict between China and Taiwan and the more recent episode prompted by *Ten Conditions of Love* demonstrate viable application, and have provided an introductory template for further case studies and more detailed analysis.

There is obviously much room and need for further study, which could take a variety of tangents. A wider assessment of the hacktivist activity occurring in aid of the various discursive struggles taking place in the region, such as those of Tibet and the Falun Gong, would build a more thorough picture of this particular regional "public of publics," and would shed light on these numerous other struggles and their battles for discursive stabilisation. An assessment of the capability of each counterpublic to engage in hacktivist expression would be a worthwhile goal, as it would appear that Chinese hackers by far outnumber their adversaries in both numbers and strength. The repercussions of this, viewed in combination with the ability of each counterpublic to mobilise various other forms of publicity, such as offline activism, would generate a partial map of the discursive power system of the region. Assessing the "success" of particular mobilisations or actions is a more complicated task, and is likely best tackled through further case studies.

There is also a clear need to assess hacktions in aid of a variety of other discursive goals, and to investigate each of Samuels' categories through a neo-Habermasian lens. The possibilities here are numerous, as is the potential and need for less transgressive but equally creative forms of political communication to be considered. Activities including but not limited to traditional street activism, political graffiti, culture jamming, and other *avant-garde* forms of political communication and protest warrant analysis and exploration, in order to broaden the repertoire of legitimate public sphere discursive forms, and facilitate the critical and inclusive political function that is integral to the concept. Houghton and Breindl (2010, forthcoming) have begun to investigate these further applications with an assessment of the Internet blackout campaigns in New Zealand and France against proposed "three-strikes" copyright legislation, but there is much room left for exploration.

The wider project of continually re-contextualising and refining the concept of the public sphere to allow for inclusivity and the recognition of difference and power is one that cannot afford to be neglected. In an increasingly interactive global society, riddled with networks of both power and resistance, the need for critical tools with which to understand political interactions is great, and "public sphere theory is in principle an important critical-conceptual response that should be reconstructed rather than jettisoned, if possible" (Fraser, 2006, p. 2). It is indeed possible, and the public sphere ideal, if appropriately reconstructed, continues to provide new ways for us to comprehend the global political society we participate in, in increasingly creative and innovative ways.

Notes

1. "Haxorz" is the commonly used "leetspeak" spelling of "Hackers." It may also be rendered "Hax0rz." See Breindl & Gustafsson, this volume, for a more thorough introduction to leetspeak.

2. While this genealogy orders each "generation" of hackers by chronological emergence, it should be understood that it is intended as a guide to understanding rather than a strict typology with no overlaps between categories and dates—hacking of all types continues today, and each new generation should be viewed as a branch extending from the central historical lineage.
3. http://www.rtmark.com/ (last accessed January 15, 2010)
4. http://theyesmen.org/ (site last accessed January 15, 2010)
5. http://www.hacktivismo.com (last accessed January 15, 2010)
6. Special thanks to Yao-Chung (Lennon) Chang (ARC Centre of Excellence in Policing and Security, Australian National University) for suggesting the case studies.
7. http://www.chinahush.com/wp-content/uploads/2009/08/20090802013.jpg

REFERENCES

Benhabib, S. (1996). Introduction: The democratic moment and the problem of difference. In S. Benhabib (Ed.), *Democracy & difference: Contesting the boundaries of the political* (pp. 3–18). Princeton, NJ: Princeton University Press.

Child, B. (2009). Chinese hackers strike again in protest over Uighur activist. *The Guardian*. Retrieved January 10, 2010, from http://www.guardian.co.uk/film/2009/sep/22/chinese-protesters-hack-website

China, Taiwan in Web hacking "war." (1999). *Greenspun*. Retrieved January 6, 2010, from http://www.greenspun.com/bboard/q-and-a-fetch-msg.tcl?msg_id=001Dzb

Costanza-Chock, S. (2001). Mapping the repertoire of electronic contention. Paper presented at the annual meeting of the IAMCR, Budapest, Hungary.

Cyberterrorism: A look into the future. (2009). *Infosecurity magazine*. Retrieved January 4, 2010, from http://www.infosecurity-magazine.com/view/5217/cyberterrorism-a-look-into-the-future/

Dahlberg, L. (2005). The Habermasian public sphere: Taking difference seriously? *Theory and Society, 34*, 111–136.

Dahlberg, L. (2007). The internet, deliberative democracy, and power: Radicalising the public sphere. *International Journal of Media and Cultural Politics, 3(1)*, 47–64.

Dahlgren, P. (1991). Introduction. In P. Dahlgren & C. Sparks (Eds.), *Communication and citizenship: Journalism and the public sphere in the new media age* (pp. 1–28). London and New York: Routledge.

Dahlgren, P. (2007). Civic identity and net activism: The frame of radical democracy. In L. Dahlberg & E. Siapera (Eds.), *Radical democracy and the internet* (pp. 55–72). Hampshire & New York: Palgrave MacMillan.

Denning, D. E. (2001). Activism, hacktivism and cyberterrorism: The internet as a tool for influencing foreign policy. In J. Arquilla & D. Ronfeldt (Eds.), *Networks and netwars: The future of terror, crime, and militancy* (pp. 239–288). Santa Monica, Arlington & Pittsburgh: RAND.

Downey, J., & Fenton, N. (2003). New media, counter publicity and the public sphere. *New Media and Society, 5(2)*, 185–202.

Dryzek, J. (2000). *Deliberative democracy and beyond: Liberals, critics, contestations*. Oxford, UK: Oxford University Press.

Dryzek, J. (2006). *Deliberative global politics*. Cambridge & Malden, MA: Polity Press.

Fraser, N. (1992). Rethinking the public sphere: A contribution to the critique of actually existing democracy. In C. Calhoun (Ed.), *Habermas and the public sphere* (pp. 109–142). Cambridge and London: MIT Press.

Fraser, N. (2006, August 27). Transnationalizing the public sphere. Retrieved February 8, 2007, from

http://www.republicart.net/disc/publicum/fraserol_en.pdf
Grey Goose 2 ties Kremlin more closely to Georgia cyber-attacks. (2009). *Infosecurity magazine*. Retrieved January 4, 2010, from http://www.infosecurity-magazine.com/view/762/grey-goose-2-ties-kremlin-more-closely-to-georgia-cyberattacks/
Habermas, J. (1989). *The structural transformation of the public sphere*. Cambridge, MA: MIT Press.
Houghton, T. J., & Breindl, Y. (2010, June 22–26). *Techno-political activism as counterpublic spheres: Discursive networking within deliberative transnational politics?* Paper presented at the International Communication Association's 60th Annual Conference, Singapore.
Jia, C. (2009). Hacker attacks website over Kadeer film. In *China Daily*. Retrieved January 4, 2010, from http://www.chinadaily.com.cn/china/2009-09/22/content_8719448.htm
Jordan, T., & Taylor, P. (1998). A sociology of hackers. *The Sociological Review*, 46(4), 757–780.
Jordan, T., & Taylor, P. A. (2004). *Hacktivism and cyberwars: Rebels with a cause?* London and New York: Routledge.
Keane, J. (2000). Structural transformations of the public sphere. In K. L. Hacker & J. van Dijk (Eds.), *Digital democracy: Issues of theory and practice* (pp. 70–89). London, Thousand Oaks & New Delhi: Sage Publications.
Laclau, E., & Mouffe, C. (1985). *Hegemony and socialist strategy*. London: Verso.
Levy, S. (1984). *Hackers: Heroes of the computer revolution*. Garden City and New York: Anchor Press/Doubleday.
McLaughlin, L. (2004). Feminism and the political economy of transnational public space. In N. Crossley & L. M. Roberts (Eds.), *After Habermas: New perspectives on the public sphere* (pp. 156–176). Oxford, UK: Blackwell.
Mouffe, C. (1993). *The return of the political*. London & New York: Verso.
Mouffe, C. (2000a). Deliberative democracy or agonistic pluralism. *Reihe Politikwissenschaft / Political Science*, (Series 72). Retrieved February 8, 2007, from www.users/unimi.it/dikeius/pw_72.pdf
Mouffe, C. (2000b). *The democratic paradox*. London: Verso.
Mouffe, C. (2005). *On the political*. London & New York: Routledge.
Nissenbaum, H. (2004). Hackers and the contested ontology of cyberspace. *New Media & Society*, 6(2), 195–217.
President Lee Teng-hui's responses to questions submitted by Deutsche Welle. (1999). *Federation of American Scientists*. Retrieved January 6, 2010, from http://www.fas.org/news/taiwan/1999/0709.htm
Ryan, M. P. (1992). Gender and public access: Women's politics in Nineteenth Century America. In C. Calhoun (Ed.), *Habermas and the public sphere* (pp. 259–287). Cambridge, MA & London: The MIT Press.
Samuel, A. W. (2004). *Hacktivism and the future of political participation*. Unpublished doctoral thesis, Harvard University, Cambridge, MA.
Schneier, B. (2008). Chinese cyber attacks. *Schneier on Security*. Retrieved January 6, 2010, from http://www.schneier.com/blog/archives/2008/07/chinese_cyber_a.html
Shapiro, I. (2004). Power and democracy. In F. Engelstad & Ø. Østerud (Eds.), *Power and democracy: Critical interventions* (pp. 11–32). Aldershot & Burlington UK: Ashgate.
Spokesman on Lee Teng-Hui's separatist malice. (1999). *China Embassy*. Retrieved January 6, 2010, from http://www.china-embassy.org/eng/zt/twwt/t36718.htm
Taylor, P. A. (2001). Hacktivism: In search of lost ethics? In D. Wall (Ed.), *Crime and the internet* (pp. 59–73). London and New York: Routledge.
Tran, M. (2009). Chinese hack Melbourne film festival website to protest at Uighur documentary. *The Guardian*. Retrieved January 10, 2010, from http://www.guardian.co.uk/world/2009/jul/26/rebiya-

kadeer-melbourne-film-china

Turkle, S. (1984). *The second self: Computers and the human spirit*. London: Granada.

Vegh, S. (2003). *Hacking for democracy: A study of the internet as a political force and its representation in the mainstream media*. Unpublished doctoral thesis, University of Maryland, College Park, MD.

Why Taiwan must show "The 10 Conditions of Love." (2009). *Taiwan News*. Retrieved January 10, 2010, from http://www.etaiwannews.com/etn/news_content.php?id=1024860&lang=eng_news&cate_img=46.jpg&cate_rss=news_Editorial

Young, I. (2001). Activist challenges to deliberative democracy. *Political Theory, 29*(5), 670–690.

PERSPECTIVE 2:

CYBERCRIME: A NEW CHALLENGE FOR LEGISLATION AND INTERNATIONAL NEGOTIATION

Yao-Chung Chang

No doubt the internet should be considered one of the greatest inventions to affect human society. However, the "network of networks" is also the proverbial "doubled-edge sword." Along with the conveniences it may bring to social, economical, and political aspects of our lives comes the inconvenience of computer crime. The spread of computer network access to the domestic arena can be seen as a gateway to cyber criminals and to "cyberdeviant" entrepreneurs. As Lessig (1999, p. 39) argues: "[t]he internet was built for research, not commerce. Its protocols [are] open and unsecured; it was not designed to hide. Data transmitted over this net [can] easily be intercepted and stolen; confidential data [can] not easily be protected." In line with this, hi-tech crime becomes a serious issue. Hackers have evolved from simply showing off their superior computing abilities, to (in some instances) stealing data or illegally obtaining financial benefits.

Governments now face the extensive problem of security and data protection, as do private companies—especially those with large personal data files, such as banks and telecommunication companies. Whenever companies and governments store, process, or transmit their customers' personal or classified data, business records, and other sensitive data in computers or through the internet, they become a potential target for criminals. For this reason, the problem of how to build a secure computing environment for companies and governments has become a worldwide concern.

The Emergence of Cybercrime

In terms of the neutrality of technology, the development of the internet not only provides opportunities for enterprises and governments, but it also provides opportunities for criminals. As Lessig (1999) argues, the internet was not originally built for commercial use, thus security was not emphasised at the beginning of its development. While the openness of the internet facilitates the open and extensive transmission of information and data, it also facilitates the activities of criminals working to exploit that openness. Traditional criminal activities have moved online, and perhaps more importantly, the internet has created new avenues for unprecedented levels and kinds of deviance and crime.

A Definition of Cybercrime

Cybercrime, which is sometimes called "computer crime," "computer-related crime," "hi-tech crime," "technology-enabled crime," "e-crime," or "cyberspace crime," still has no precise and clear academic definition. Even the Council of Europe's Convention of Cybercrime (the first convention related to Cybercrime, which was adopted by the Council of Europe on 8 November 2001 and ratified by 15 States as of 2 September 2006) has not adopted a uniform definition.

According to Wall (2007, p. 10), the word "cybercrime" tends to be used "metaphorically and emotively rather than scientifically or legally." Just as the term "white collar crime" has been used for about 50 years, academia uses these terms to "delimit the scope of computer-related misconduct" (Smith, Grabosky, & Urbas, 2004, p. 5). It is used to describe criminal offences specifically related to computers and telecommunications, or the crimes that take place within cyberspace (Smith et al., 2004; Wall, 2007).

Although there is no clear definition of cybercrime, it can be classified into three general forms according to legislation and/or the common law developed in the courts (Grabosky, 2007b):

1. Crimes where the computer is used as the *instrument* of crime, such as phishing, producing and disseminating child pornography;
2. Crimes where the computer is the *target* of crime, such as denial of service attacks; and
3. Crimes where the computer is *incidental* to the offence, such as maintaining records of criminal transactions such as money laundering and drug dealing.

Grabosky (2007a) notes that although the typology is useful, there is significant overlap in crimes where the computer is both the target and the instrument.

The Evolution of Hackers

From the 1950s to the 1980s, cybercrime was not particularly popular because of the high level of skills required to become a "hacker." Most hackers of that era were driven by technological curiosity and by the thrill of gaining knowledge and beating the system (Choo, Smith, & McCusker, 2007; Levy, 1984; Taylor, 2005). Most of them obeyed three key core elements of "hacking ethics" described by Taylor (2005): the ingenious use of technology, the tendency to subversively reverse-engineer technology, and the desire to explore systems. Generally, most hacking during that period was based on the concept of promoting rather than hindering human agencies (Choo et al., 2007), and was thus not "criminal." This "genealogy" and hacking's continued and tangential evolution has been discussed in more detail by Houghton (Perspective 1).

It was only in the mid 1990s that cybercrime came to be considered a social problem. Hacker activity expanded from the relatively amicable behaviour of the 1980s to include the malicious cyber criminal behaviour of the mid to late 1990s (Jewkes, 2003; Williams, 2006). Unlike early hackers, some modern hackers, or "crackers," are motivated more by financial gain, revenge, or politics (Choo et al., 2007) even though their hacking methods might be the same as those employed by the earlier "ethical hackers."

When criminal hackers discover vulnerabilities or breaches of websites or network protocols, they will not tell the owner to patch it. Rather, they will hack into the system or network, avoiding detection from the gatekeepers or the owners of the networks or computers, to procure any trade secrets, classified documents, or other useful information contained therein.

They are also able to turn the hacked computer into a zombie so that it hacks other computers or networks. Sometimes, through the help of insiders or the use of "social engineering" skills to acquire names and passwords to access networks or websites, hackers can also perform cyber-espionage (Wall, 2007). Agencies trusted with large personal data files, or those who rely heavily on the internet to perform their services (such as government agencies and banks), can easily become the victims of cyber criminals (Symantec, 2006). Hackers can often gain considerably more advantage by hacking into organisational systems than into personal computers.

The De-skilling of Cybercrime

In addition to the evolution of hackers, the commercialisation of hacking software and skills, and the "de-skilling and re-skilling of criminal labour" (Wall, 2007, p. 43) have increased the challenge of cybercrime control. Initially, anyone who wanted to be a cyber criminal had to be well-trained. One needed professional knowledge about computers and networks to be able to hack into other computers and networks to steal personal data or to turn the computer into a zombie.

However, the "automation of the process" makes it possible for anyone to become a hacker. One does not need skill, just the knowledge of where to procure and how to use appropriate software, and the malicious code and software will run automatically (Wall, 2007). That is, anyone who knows how to use the internet can become a "script-kiddie" with the help of software, or by hiring a true hacker.

This has attracted the attention of governments and academia. Governments are establishing contingent strategies to deal with cybercriminal attacks, particularly to deal with the threat of cyber-terrorism, and are examining their laws and regulations to tackle the phenomenon of cybercrime. Much research focuses on this area as well.

THE INADEQUACY OF LAWS AND REGULATION TO TACKLE CYBERCRIME

The Limitations of Law and Regulation

Since cybercrime has become a serious problem, the question of how to deter cybercriminals is widely discussed. Although some people are strongly against government legislative intervention to regulate cyberspace (Barlow, 1996; Goldsmith & Wu, 2006), legislation may be the only way ahead. In practice, it is the way most countries are seeking to combat cybercrime. However, there is much evidence that laws and regulations are largely ineffective. Cairncross (1997, p. 4) argues that "[u]ndoubtedly, governments will find that national legislation is no longer adequate to regulate a global flow of information, even if some of that information is criminal or subversive."

Grabosky, Smith, & Dempsey (2001) and Katyal (2003) also judge that laws and regulations will not be the primary means to secure cyberspace, even where they are kept up to date with the development of technology. The character of cross-border crime, which limits a government's capacity to control behaviour in cyberspace, and the lack of law enforcement capacity (well trained investigators and services) to investigate crimes happening in cyberspace, contribute to the limited effectiveness of laws and regulations (Grabosky et al., 2001). Although the "cat and mouse game" still exists between cybercriminals and law enforcement agencies, not every investigator has the skills necessary to investigate cybercrime. Furthermore, as hackers exploit new skills or techniques to commit crime, investigators must constantly catch up with ever-developing skills and techniques (Wall, 2007).

Laws and regulations also have limited effect because most computers and networks are privately owned. The National Infrastructure Advisory Council of the United States identifies the reason why governments might not be able to secure cyberspace alone:

> Most critical infrastructures, and the cyberspace on which they rely, are privately owned and operated. The technologies that create and support cyberspace evolve rapidly from private-sector and academic innovation. Government alone cannot sufficiently secure cyberspace. (2003)

The blurred lines between public networks and private networks, between private networks, and even between national networks are making the world borderless (Katyal, 2003). Information can easily flow anywhere, as can malicious code. Furthermore, although the commercialisation of software brings convenience to the world, it also attracts hackers. Once hackers discover vulnerabilities or breaches in commercialised software, such as Microsoft Windows Operating Systems, they can use these "bugs" to hack into both private and public sector systems alike.

Furthermore, the nonreporting of cybercrime results in a sub-optimal capability to address problems and risks widespread contagion. If the victim reports the incident properly, it not only helps software companies to patch vulnerabilities quickly, but also brings the problem to the attention of other companies and government agencies. The development of "pre-warning" and "information sharing" systems has become increasingly important (Choo et al., 2007; National Infrastructure Advisory Council, 2003; Wall, 2007). Nevertheless, most companies are not willing to let others know that they have suffered an information security problem for the fear of losing public confidence and market share. Managing this conflict will be central to promoting information sharing systems (Chang & Wu, 2008; Dacey, 2004).

The Cross-Border Problem

As mentioned, government agencies and banks are the main targets for hackers. Despite their desire for secrecy, news of hackers breaking into banks or government agencies online services to steal personal data, business secrets, identities and passwords, classified data, and other data is a frequent occurrence.

In 2007, there was a dramatic attack on Estonia, which originated from Russia. This so-called "cyberwar" hit government and bank websites, including the Estonian presidency, parliament, almost all of the government ministries, political parties, organisations, and big banks (Traynor, 2007). There is also evidence of espionage activities carried out by Chinese hackers on the assets of other countries. China has been accused of supporting hackers who penetrated the highly classified government computer networks of the United States, the United Kingdom, Germany, France, Australia, and Taiwan (Evans, 2007; Shih, 2007; Walters, 2007). Irrespective of whether governments officially support such hacking activities or not (Cody, 2007; Lu, 2007), these events indicate that as governments and banks rely increasingly on the internet, they also increase their risk of becoming the targets of cybercriminals.

Cairncross (1997, p. 3) states "the death of distance loosens the grip of geography... Barriers and borders will break down. The horizontal bonds among people doing the same job or speaking the same language in different parts will be stronger." However, Katyal (2003, p. 180) notes that countries will also find it increasingly difficult to enforce their national laws against external subjects that are offensive or harmful to local taste or culture. It is the internet that allows hackers to plan hacking activities in one country and to hack into others.

Therefore, even if cybercriminals are caught, the jurisdiction problem will be challenging. In other words, inconsistencies between laws and regulations in different countries may make it difficult for them to cooperate in the investigation of crimes. In addition to combating this jurisdictional inconsistency, the establishment of trust and cooperation between countries is also an important issue in combating cross-border crime.

Conventions and agreements have been developed by international organisations such as the Council of Europe, in the interest of encouraging inter-jurisdictional harmonisation of cyber-laws and regulations, and to build up cooperation among nations in combating cybercrime. Early in 1986, the Organisation for Economic Cooperation and Development (OECD) published a study on "Computer-related Crime: Analysis of Legal Policy." The pamphlet discusses developments in OECD member countries aimed at combating the new and growing phenomena of computer-related crimes, and emphasizes the importance of developing common laws (p. 64):

> It is important to develop common approaches for penal law and for procedural law in order to protect international data networks, to enable the functioning of international instruments of cooperation in criminal matters and to guarantee that evidence gathered in one country is admissible in court in another country.

In 2001, the United Nations also declared a resolution on combating cybercrime, expressing concern at the effect of technological advancement on criminal activities. Moreover, the resolution called upon its members to note the "Convention on Cybercrime" drafted by the Council of Europe, which is now the most accepted convention on cybercrime (United Nations, 2001). As of now, despite some criticisms, most cyber laws and regulations have been made consistent with the spirit of the "Convention on Cybercrime."

Concluding Remarks

Many hackers today are more profit-oriented than the early hackers who simply hacked for fun. They target companies or government agencies which hold sensitive data or huge personal databases, and who rely extensively on the internet for their services. When these companies and government agencies are hacked, significant harm can occur. As such, the establishment of a better way of preventing companies and government agencies from being hacked is crucial, and is an issue that cybercrime research should focus on.

Moreover, the consistency of laws and regulations among nations is important to combating cross-border crimes like cybercrime. Establishing suitable laws and regulations consistent with international conventions and treaties is the most effective way for countries to act together to tackle cybercrime.

Nevertheless, laws and regulations are only useful as a deterrent or to penalise cybercriminals when they are caught; they cannot prevent the damage from occurring. Hence, the possibility of establishing "pre-warning" and "information sharing" systems could compensate for the limited effectiveness of laws and regulations. In a world where public and private sector data is increasingly moving online, establishing the security of data is an essential undertaking, and one that requires ongoing attention.

References

Barlow, J. P. (1996). A declaration of independence for cyberspace. Retrieved 28th Sept, 2007, from http://homes.eff.org/~barlow/Declaration-Final.html

Cairncross, F. (1997). *The death of the distance: How the communications revolution will change our lives.* Boston, MA: Harvard Business School Press.

Chang, Y. C., & Wu, J. (2008). Study on information security incidents reporting and information sharing mechanism in Taiwan—From the viewpoints of the U.S. experience. *Science and Technology Law Review, 20*(8), 39–61.

Choo, K.-K. R., Smith, R. G., & McCusker, R. (2007). *Future directions in technology-enabled crime: 2007–2009.* Retrieved 15th December 2009 from http://www.aic.gov.au.

Cody, E. (2007). Chinese official accuses nations of hacking. Retrieved 26th September, 2007, from http://www.washingtonpost.com/wp-dyn/content/article/2007/09/12/AR2007091200791.html

Dacey, R. F. (2004). *Critical infrastructure protection—Establishing effective information sharing with infrastructure sectors.* Washington, DC: General Accounting Office.

Evans, M. (2007). China "tops list" of cyber-hackers seeking UK government secrets. Retrieved 26th September, 2007, from http://www.timesonline.co.uk/tol/news/world/asia/article2393979.ece

Goldsmith, J. T., & Wu, T. (2006). *Who controls the internet? Illusions of a borderless world.* New York, NY: Oxford University Press.

Grabosky, P. (2007a). *Electronic crime.* Upper Saddle River, NJ: Pearson.

Grabosky, P. (2007b). Requirements of prosecution services to deal with cyber crime. *Crime, Law and Social Change, 47,* 201–223.

Grabosky, P., Smith, R., & Dempsey, G. (2001). *Electronic theft: Unlawful acquisition in cyberspace.* Cambridge, MA: Cambridge University Press.

Jewkes, Y. (2003). *Dot.cons: Crime, deviance and identity on the internet.* Cullompton, UK: Willian.

Katyal, N. K. (2003). Digital architecture as crime control. *Yale Law Journal, 112*(8), 2261–2289.

Lessig, L. (1999). *Code and other laws of cyberspace.* New York: Basic Books.

Levy, S. (1984). *Hackers: Heroes of the computer revolution.* New York: Bantam Doubleday Dell.

Lu, A. (2007). Premier Wen: China opposes hacker activity. *Xinhuanet.* Retrieved 21st July 2009, from http://news.xinhuanet.com/english/2007-08/27/content_6611437.htm

National Infrastructure Advisory Council. (2003). *The national strategy to secure cyberspace.* Retrieved 26th December 2007, from http://www.whitehouse.gov/pcipb/cyberspace_strategy.pdf

OECD. (1986). *Computer-related crime: Analysis of legal policy.* Paris: OECD.

Shih, H. C. (2007). GIO prohibits work from home after hacker attack. Retrieved 23th September, 2007, from http://www.taipeitimes.com/News/taiwan/archives/2007/05/26/2003362509

Smith, R., Grabosky, P., & Urbas, G. (2004). *Cyber criminals on trial.* Cambridge, MA: Cambridge University Press.

Symantec. (2006). *Symantec internet security threat report X: Trends for January—June 06.* Cambridge, MA: Symantec Corporation.

Taylor, P. A. (2005). From hackers to hacktivists: Speed bumps on the global superhighway. *New Media Society, 7*(5), 625–646.

Traynor, I. (2007). Russia accused of unleashing cyberwar to disable Estonia. *The Guardian.* Retrieved 23rd September, 2007, from http://www.guardian.co.uk/russia/article/0,,2081438,00.html

United Nations. (2001). *Resolution adopted by the General Assembly on combating the criminal misuse of information technologies (A/RES/55/63).* Retrieved 23rd September, 2007, from: http://www.unodc.org/pdf/crime/a_res_56/121e.pdf

Wall, D. S. (2007). *Cybercrime.* Cambridge, UK: Polity.

Walters, P. (2007). China's cyber raid on agencies. *The Australian.* Retrieved 23rd September, 2007, from: http://www.theaustralian.news.com.au/story/0,25197,22404058-5001561,00.html

Williams, M. (2006). *Virtually criminal—Crime, deviance and regulation online.* New York: Routledge.

CHAPTER ELEVEN

ICTs and the Green Economy

US and Chinese Policy in the 21st Century

DANIEL ARAYA, JIN SHANG, & JINGFANG LIU

The growing consensus around climate change has led many analysts to champion information technology as a foundation to a green economy. Reducing global warming and enhancing resource management are strong reasons to take the idea of a green economy seriously. Many countries today are aggressively pursuing green innovation strategies because of the potential social and economic benefits associated with harnessing clean energy. Much as information and communications technologies (ICTs) have underwritten globalization and reshaped industrial societies, they are transforming the structure and practices of green innovation. ICTs are no silver bullet, but they are very likely a major key to developing a clean energy economy. In this chapter we explore the policy implications of green innovation in the context of the two largest economies in the world, the U.S. and China. Looking critically at U.S. and Chinese economic policy, we consider the opportunities and challenges facing these two economic giants. We begin with U.S. policy in relation to the green economy, and then turn to Chinese policy. What should be the green economy strategies for the U.S. and China going forward? How can they work cooperatively to resolve the growing economic and political challenges around climate change?

The Green Economy and U.S. Policy in the 21st Century

It has become increasingly clear that we are moving beyond the closed loop of industrial civilization. Aggressive attempts to curb ozone depleting chemicals and cut green house gas emissions have become commonplace in contemporary public policy discourse. Growing concern about climate change and the impact of carbon dioxide emissions has made green innovation and a green economy a central feature of international debate. At the policy level, the definition of a "green economy" remains in flux, but is linked to the elimination of fossil fuels and a long-term shift from highly polluting industrial industries to low-carbon, low-waste industries. This includes technological innovation, renewable energy generation (wind, solar, geothermal, and biomass) and energy conservation.

Many government programs around the world today focused on environmental sustainability are exploring technology initiatives for reducing green house gases. Denmark's *Action Plan for Green IT* established by the Ministry of Science, Technology, and Innovation, for example, and Japan's *Green IT Initiative* established by the Ministry of Economy, Trade, and Industry, provide strong models of cutting edge green innovation policy (Castro, 2009). Although there is no fixed definition of "Green ICTs," Green IT or Green Computing is often defined as the practice of using computers and telecommunications in a way that maximizes positive environmental benefit and minimizes environmental pollution (Worthington, 2009, p.16). This includes energy efficiency in ICT equipment and the use of recyclable materials in the production of ICT-related goods. In the U.S., two national agencies have each initiated green ICT-related measures, the U.S. Department of Energy (DOE)[1] and the U.S. Environmental Protection Agency (EPA).[2]

Technology may be essential to green innovation but strategic public policy is critical. Evidence from climate science points to the need for a large-scale coordinated response to climate change, including a broad portfolio of active technologies and regulatory management. Reducing global warming and enhancing resource management are strong reasons to take the idea of green ICTs and a green economy seriously (Castro, 2009). The limitations of past U.S. policy, however (particularly when compared to countries at the leading edge of green innovation) suggest that the U.S. has a long way to go. Consider the data: 85 percent of America's energy comes from fossil fuels (petroleum, coal, and natural gas). While America produces 10 percent of the world's oil, it consumes one quarter of the world's total supply. 70 percent of that oil is imported from abroad (with estimates of $475 billion on imported oil in 2008 alone). The hard reality is that America's high-tech society is built on top of a fossil fuel economy of oil, coal, and natural gas. With finite supplies of carbon energy and global warming concerns, this situation will have to change dramatically.

While environmental policy in the U.S. has mainly focused on educating (and sometimes miseducating) the private sector about its environmental impact, what the U.S. has lacked most is a robust public policy for coordinating a coherent green strategy. The strength of countries at the leading edge of green innovation, for example, is anchored in their capacity to leverage government oversight in the rapid coordination of public-private partnerships. Lacking a coordinated and vigorous energy policy, the U.S. has not been as effective as other advanced industrial countries.

Challenges to the U.S. Economy

Where does the U.S. economy stand today? Over the past 20 years, structural weakness in the U.S. economy has become considerable. Masked by a succession of economic bubbles (the dotcom bubble, the low interest bubble, the real estate bubble, and most recently the subprime mortgage crisis), the U.S. economy has become structurally reliant on foreign capital to finance itself. Weak manufacturing has led to a massive trade deficit and a growing external debt. As American industrial production has shrank, domestic consumption has soared.

While the U.S. economy is heavily dependent on consumer spending, American consumers are now deeply in debt. Increasingly reliant on foreign creditors, the U.S. appears to be a global hegemon in decline. According to Goldman Sachs, the U.S. share of global gross domestic product fell to 27.7 percent in 2006 from 31 percent in 2000. Meanwhile, the share of global gross domestic product for the BRIC countries (Brazil, Russia, India, and China) rose to 11 percent from 7.8 percent (Gross, 2007). China alone accounts for 5.4 percent. In 2007 the BRICs' contribution to global growth was in fact greater than the U.S. Even adjusting for the differential power of currencies, growth in the U.S. has lagged behind global growth for the past 10 years.

The hard reality is that the American economy lacks the productive capacity to rejuvenate itself. Many now believe that the U.S. needs to develop a new economic engine. Van Jones (2008), for example, has proposed a "green new deal" to transform the U.S. around alternative energy. Insulating homes and businesses alone he suggests, could produce a service economy of low and medium-skill jobs that could help to revitalize the U.S.

A New Energy Paradigm: Renewing the U.S. Economy

The rising call for green economy jobs offers significant promise both in terms of innovation and mass employment. Beyond fossil fuels, the continental U.S. has enormous assets in renewable wind and solar energy. Underlying any strategy for renewable energy, however, is the issue of the storage and transmission. Most analysts agree that the transmission and distribution of electricity is a major key to developing a green economy. America's aging electricity grid, however, is in critical need of renewal. Changes to the U.S. electricity grid, for example, would likely mean introducing next generation digital technologies.

The difference between traditional grids and the potential of digital grids is considerable. While traditional grids simply distribute energy, next generation *Smart Grids* have the capacity for two-way energy transmission. A two-way power flow from distributed renewables like solar and wind, for example, would mean that consumers could sell excess electricity back to the grid. Directly connecting suppliers (utilities) to customers, while at the same time transforming this into a bi-directional relationship, would completely transform the U.S. energy infrastructure .

Using smart grids, U.S. energy production could become augmented by clusters of locally distributed energy resources (DER)[3]. Using a combination of advanced two-way technologies and applications, energy generation in the U.S. could look increasingly like the Internet. Policy related organizations like GridWise (2009) for example, argue that smart grids can transform the production, storage, and use of energy, just as the Internet revolutionized the production and use of information and communications. While DER systems are small-scale power-generation technologies (typically in the range of 3–10,000 kW), they can be combined to form dense clusters of networked energy. Since most forms of renewables (ocean tide, wind, solar) are inherently variable or intermittent, smart grids can balance sudden drops in electrical generation by adjusting storage or consumption itself. Linking consumers and producers into a single national energy network could have huge potential for green innovation.

U.S. Policy and the Green Economy

The recent rush of stimulus spending around the world includes significant investments in the green economy. Green innovation and the development of green technologies are seen by many policymakers as critical to the future prosperity of advanced economies. This is clearly the view of the Obama administration as well. The U.S. administration's stimulus bill has dedicated U.S.$71 billion to clean energy funding, with an additional U.S.$20 billion for loan guarantees and tax incentives to support clean energy projects. The administration has proposed investing in a *Unified National Smart Grid* linking all of the nation's local electrical networks that have been upgraded to smart grids. High-capacity transmission would span the country, providing linkages to local electric utilities and distantly located bulk power generation facilities.

Using advanced, high-voltage lines, America's Unified Smart Grid is envisioned as efficiently moving electricity across vast geographic distances with minimal loss-es. Long distance interconnections are not new (a 5 GW 800 KV system is currently being constructed along the southern provinces of China). What is new, however, is the network architecture of the Unified Smart Grid. The Unified Smart Grid would not simply be a collection of point-to-point interconnections between regional systems. Rather, it would be a two-way electrical transmission network, enabling access points to function as virtual power generators or Grid energy storage facilities. Smart grid applications, for example, would improve the ability of electricity producers and consumers to communicate with one another and make decisions about how and when to produce and consume energy. Smart-grid operators could move energy provided by wind or solar power plants from one side of the country to the other. Combining solar power from the South with wind power from the Northeast, the U.S. would be able to use its varied geography in a highly coordinated way.

Smart Grids not only introduce additional layers of advanced technology, they also bridge adjacent markets in architectural design, smart home appliances, wireless networks, auto production, and traditional utilities. At the most basic level, however, smart grids would mean a wider array of customized energy choices, commercial technologies and services, distributed intelligence, and clean power. Deploying a unified energy network would mean consumers could also become producers, collaborating together around open innovation. This could mean a new democratic platform for prosumer (producer-consumer) collaboration (von Hippel, 2005). Bauwens (2009), for example, argues that these networking systems underlie the gestation of new socioeconomic infrastructure in desperate need of a supportive institutional and governmental framework. Beyond the command-and-control systems characteristic of mass industrial society, the green economy would likely be coordinated by the collaborative logic of peer-to-peer networks.

Global Shift

Beyond the U.S. and other advanced economies, however, there is a global crisis looming around energy consumption in emerging economies. China is predicted to become the largest economy in the world by the middle of this century. While China and India only accounted for 10 percent of the world's energy consumption in 1990, and 19 percent in 2006, they are projected to account for 28 percent in 2030[4] (Energy Information Administration, 2009). As Victor and Yueh (2010) argue, the "era of growing demand for oil and other fossil fuels in the industrialized countries is over; most of the future growth in demand will come from the energy market countries, notably China and India (p. 62)." Rising demand for energy in developing countries is putting unprecedented pressure on the global energy system. Beyond economic nationalism, there is a critical need for new collaborations in green policy and green innovation. Consider for example, the recent stimulus spending to rejuvenate the global economy:

> The problem is most obvious regarding the "green" part of the $2.5 trillion that is being spent globally to stimulate the world economy. The U.S. and China alone are spending $1.5 trillion, including a large fraction on energy projects. South Korea has devoted 85 percent of its stimulus package to green investments, promoting energy efficiency and low-emissions power plants. The British government has set aside hundreds of millions of pounds to support research and development in green industries. Coordination is needed, however, because the market for green-energy technology is global; ideas promoted in one country can quickly spread to the rest of the world through the marketplace. For example, U.S. spending on renewable sources of energy can invigorate U.S., Chinese, and European firms that supply solar cells and wind turbines, boosting all three economies at the same time. And Chinese spending on new power grids can benefit the Western companies, as well as the Chinese ones, that develop the requisite technology. (Victor and Yueh, p. 70)

The bottom line is that the green economy is different from other markets and will require political coordination at the highest levels. Part of the answer to this lies in commercial markets themselves—free competition should bring the best green technologies to the forefront. But beyond commercial markets, there is the need for coordination in public policy to underwrite green innovation and negotiate the frictions between competing nation-states. Although the growing list of energy consuming countries around the world is long, priority belongs to the world's largest economies, particularly the U.S. and China. Given their dominant roles as the world's largest energy consumers, the U.S. and China must play a significant role in nurturing the green economy. In order to appreciate the scope of this challenge, it is important to understand China's current economic trajectory.

The global economy today is still largely based on industrial production. For this reason, international trade in industrial products is still viewed as one of the major indicators of global GDP growth. As China continues to modernize and build consumer-driven markets, it is absorbing an increasing share of the world's energy and resources. Behind the story of China's unprecedented economic growth is an increasing consumption of natural resources, especially energy.

China has become one of the largest greenhouse gas emitters in the world today, and "energy consumption in China is expected to rise significantly as the country aims to quadruple its gross domestic product by 2020" (Jia, 2004, para. 5). Interestingly, China will likely surpass the U.S. in smart grid investments this year, investing $7.3 billion dollars in smart grid technology through stimulus loans, grants, and tax credits (compared to $7.1 billion by the U.S.) (Bhanoo, 2010). In order to cope with growing environmental problems and improve the sustainability of China's economic development, the Chinese government has introduced policies to support "environment construction." According to Liu Zhiquan, Deputy Director General of Department of Technology Standards, Ministry of Environmental Protection, China plans to invest 3.1 trillion Yuan (or $450 billion U.S.) in environmental protection, exceeding that of the U.S. and Japan (People's Daily Online, 2009, para. 5). As a developing country however, China is largely focused on industrialization.

For much of the 20th century, China did not substantially participate in the global economy. However, over the past 30 years the Chinese economy has undergone significant restructuring linked to commercial markets and privatization. Today, China has moved past Japan to become the world's second largest economy. In the last two decades China has overtaken the U.S. and Germany to become the world's largest exporter and it now has the world's largest foreign exchange reserves. The downside of all this economic growth is that China has also become the world's largest emitter of carbon (Mandelson, 2010).

The struggle between environmental degradation and environmental protection has been daunting for China. Early efforts at environmental protection in China can be traced to its 1972 delegation to the UN's Human Environment Conference[5] and to China's environmental protection conference in 1973. Under market liberalization, China has found the challenge of implementing environmental policy at the local level to be a major obstacle to managing environmental pollution. Over the past three decades, the rate of environmental pollution in China has significantly outpaced the country's capacity to deal with environmental protection (Economy, 2004). This is not to say that the Chinese government is taking no action on climate change. Unlike the market-based approach to environmental protection seen in the U.S., the Chinese government has been quite proactive in establishing a comprehensive national apparatus for promoting environmental protection and sustainable economic growth.

THE GREEN ECONOMY AND CHINA IN THE 21ST CENTURY

Today, the long-term policy trajectory of China is to gradually shift from "high-energy consumption" to "information processing" (Hu, 2007, para. 3). China's industrialization is only accelerating, however, and industrial development remains the foremost goal of Chinese economic policy (National Development and Reform Commission, 2006, para. 3). The good news is that Chinese policymakers are acutely aware of the need to employ information technologies in order to develop a "highly-efficient" and "low-cost" ICT infrastructure. Over the past two decades, a shift from industry to information has become a central feature of Chinese innovation policy. ICTs, it is suggested, have the long-term potential to replace material goods (dematerialization), and increase productivity and efficiency (Hilty, Seifert, & Treibert, 2005). However, the reality is that ICTs linked to globalization are boosting the production of material goods. Rather than dematerialization, what we are seeing is a third-order effect of ICTs. This suggests an underlying logic for the development of green ICT innovation that requires closer coupling to environmental protection.

There are at least four key areas of industry in which the use of IT is promoted to support Chinese industrial development. These industries include traditional industries (energy and manufacturing), agricultural industries, service industries, and industrial enterprises. Among these four areas of industry, the traditional industries are given the most attention by government because of their relative importance to China's current stage of industrialization (The State Council of China, 2008. para. 2). According to the State Council of China (2006, para. 4), energy and manufacturing are the two primary industries essential to China's long-term national development.

The problem, of course, is that China's industrialization is highly polluting. Heavily reliant on the consumption and exploitation of natural resources, the Chinese government is increasingly introducing policies to improve and even radically change their approach to economic development. Using ICTs to balance environmental protection with industrial growth is now viewed as being extremely important to China's future trajectory. The total investment in the IT sector of China's manufacturing industries, for example, grew from 28.4 billion Yuan ($4.1 billion US) in 2005 to 43.7 billion Yuan ($6.4 billion US) in 2008 (see Figure 1). Progress on the ground, however, is rather uneven due to variations in development across different regions of China.

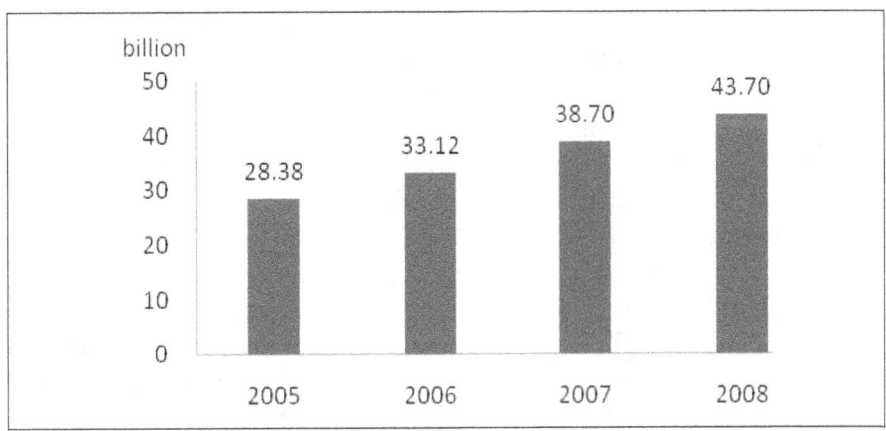

Figure 1. Growth of investment in IT sectors of China's manufacturing industries (2005–2008)
Source: Data from China's Manufacturing Industry Development Annual Report (CCID, 2009, para. 3)

Using ICTs to Manage and Control Industrial and Urban Pollution

Apart from industrial use in China, ICTs can also be widely adopted to measure and control the impact of industrial activities on the environment. As Rautenstrauch and Patig (2001) point out, one of the reasons industrial manufacturing is often seen as the cause of environmental damage is the lack of information in manufacturing industries for managing raw materials. If industrial manufacturers in China and elsewhere were to adopt specialized information systems to improve their industrial production processes, this could go a long way towards managing industrial pollution.

Rapid industrial and urban development over recent decades has left China's environment heavily damaged. Water, soil, forest, and minerals are all declining at a disconcerting rate (The World Bank Group, 2001). In response, the Chinese government has established a series of policies and regulations to control or restrain the impact of industrial pollution on the environment. Among these regulations, environmental monitoring systems and networks have been given particularly high priority. In addition, all environmental protection departments at the local level are required to construct their own advanced environmental monitoring systems to support timely decision-making. This includes environmental monitoring databases, in order to improve the capability for collecting and analyzing environmental data (Ministry of Environmental Protection, 2007, para. 3).

While the construction of China's environmental monitoring system has made good progress, one of the main obstacles to environmental protection in China is a marked disinterest from commercial enterprises on the ground. Short of economic or other incentives, many enterprises in China show little or no interest in cooperating with local authorities on environmental protection. To reverse this situation, the Chinese government has set up electronic platforms for exchange of environmental activities among Chinese enterprises. In the process of "trade exchanges," a great deal of information is processed via the Internet, which has itself improved the efficiency of "green activities" in China. These electronic exchanges include the China Beijing Environment Exchange (CBEEX), the Shanghai Environment and Energy Exchange (SEEE) and the Tianjin Climate Exchange (TCE).

One of the key aims for establishing such electronic platforms is to encourage polluting enterprises to find ways to engage in energy-saving activities while gaining commercial benefits. Such activities could assist Chinese industries to significantly cut down their CO_2 emissions. In fact, the CBEEX has recently set up the China Carbon Neutral Alliance (CCNA), which is regarded as China's first carbon-neutral organization. In return for brand value, CCNA requires that any enterprise or organization that wishes to join its network establish a timetable for carbon emission reduction procedures (China Beijing Equity Exchange [CBEX], 2010).

From Traditional Industry to Information Industry

Compared to traditional industrial industries, information and knowledge processing could offer an important new pathway for promoting green innovation and a green economy in China. Arguably, raising the proportion of "information industries" in the Chinese economy will go some way towards helping to ease the pressure of economic growth on the natural environment. Since the early 1980s, the Chinese government has made significant efforts to encourage the development and innovation of China's information industry. As early as in 1982, the State Council founded a special group called the "Computer and LSI Leadership Group," in order to boost the development of the computing industry in China.

According to a report by the Ministry of Science and Technology (MST), the electronics and communication industries had already become critical to the Chinese economy by the end of 1999 (Lu, 2002, p. 51). Over the past 10 years, China's information industry has entered a boom period, during which the Chinese government has played a decisive role in promoting the development of ICTs. In 2002, for example, the State Council Informatization Office (SCIO) enacted policy to increase government procurement and adoption of software products or services provided by local ICT enterprises in order to boost the development of the software industry and expand the international software market in China (China Software Industry Association [CSIA], 2002). As a result of these efforts, a wide variety of sectors of information industry have proliferated in China. According to a report released by the Ministry of Industry and Information Technology, the total revenue of China's information industry increased from 1.88 trillion Yuan ($275 billion U.S.) in 2003 to 5.6 trillion Yuan ($820 billon U.S.) in 2007 (Shen & Huang, 2009, para. 2).

China's emerging green ICT industry

While developing a robust information industry in China goes some way towards fostering a green economy, ICTs are no "cure-all" for achieving environmental sustainability. Due to the fast growing use of information technologies and the increasing construction of China's information infrastructure, a new kind of environmental pollution has resulted. According to Yousif (2010), the impact of ICTs on the earth's climate and its dwindling resources is a growing concern. "Recent studies found that carbon dioxide (CO_2) emissions from data centres alone surpass emissions from many individual countries ... This is in addition to the fact that much ICT equipment contains toxic substances such as lead and mercury, much of which enters the environment via the dumping of obsolete equipment" (para. 2). As one of the fastest growing segments of the Chinese economy, China is faced with an enormous problem in dealing with pollution related to ICTs (Brodkin, 2009; Chaize, 2008; Hasson, 2009)[6].

In response to the challenge of ICT related pollution, Green ICTs have become a significant area of interest in China. While still at an early stage of Green ICT development, the Chinese government has been active in embracing the concept of green technologies and engaging in the development of Green ICTs within relevant industries. While there is no formal policy to promote specific areas of Green ICTs, the Chinese government has been active in helping to facilitate platforms for business exchanges between enterprises that are interested in Green ICT cooperation. Most recently, the China Green Data Center Development and Practice Summit was held in Beijing in September 2009 under the support and direction of the Ministry of Industry and Information Technology (MIIT). In the summit, delegates from both leading enterprises and government exchanged ideas about the future development of China's Green ICT industry (Yong, 2009).

DISCUSSION: U.S. AND CHINESE POLICY IN THE 21ST CENTURY

Overall, much more needs to be done to develop clear and measurable policies to improve the performance of Green ICTs and reduce environmental pollution. The Chinese government has only recently begun to focus on policies stimulating research and development around green innovation. Beyond Green ICTs, China is investing heavily in renewable energy, including hydropower, wind, solar, biofuel, geothermal, and tidal energy. In fact, China's State Renewable Energy Medium- and Long-term Planning (SREMLP) has aimed at raising the share of China's renewable energy to 15 percent by 2020. China is now the world's largest maker of wind turbines, and the world's largest manufacturer of solar panels. At the same time, total power generation in China is on track to pass the U.S. by 2012, and most of the added capacity will still be from coal (Bradsher, 2010). Coal will represent two-thirds of China's energy capacity in 2020, and nuclear and hydropower will make up most of the rest.

THE NEED FOR COOPERATION

It is becoming abundantly clear that countries around the world must now search for ways to cooperate on comprehensive solutions to climate change. One positive example of this needed cooperation is the recent agreement between the U.S. and China to work together on clean energy policy (DOE, 2009). This joint agreement represents a strong first step in bilateral relations on green policy, including several important initiatives for incubating green technologies. These initiatives include:

1. *A U.S.-China Clean Energy Research Center* facilitating joint research and development of clean energy technologies, as well as serving as a clearinghouse to help researchers in each country. Jointly supported by public and private funding, this research center includes an investment of $150 million over 5 years.

2. *An Electric Vehicles Initiative* building on the first-ever US-China Electric Vehicle Forum in September 2009, including joint standards development, demonstration projects, technical roadmapping and public education projects.

3. *An Energy Efficiency Action Plan* supporting improved energy efficiency in buildings, industrial facilities, and consumer appliances including energy efficient building codes and rating systems, testing procedures and performance metrics for consumer products. In addition, both countries agree to convene a new U.S.-China Energy Efficiency Forum to be held annually, rotating between the two countries.

4. *A Renewable Energy Partnership* developing roadmaps for wide-spread renewable energy deployment in both countries. The Partnership will also provide technical and analytical resources to states and regions in both countries to support renewable energy deployment and facilitate state-to-state and region-to-region partnerships, including a new U.S.-China Renewable Energy Forum that will be held annually, rotating between the two countries.

5. *A cooperation program on cleaner uses of coal,* including large-scale carbon capture and storage (CCS) demonstration projects, bringing teams of U.S. and Chinese scientists and engineers together in developing clean coal and CCS technologies.

6. *A U.S.-China Shale Gas Resource Initiative* using experience gained in the U.S. to assess China's shale gas potential and promote environmentally-sustainable development of shale gas resources (including joint technical studies) in China.

7. *An Energy Cooperation Program (ECP)* leveraging private sector resources for project development work in China across a broad array of clean energy projects. This includes collaborative projects on renewable energy, smart grid, clean transportation, green building, clean coal, combined heat and power, and energy efficiency.

The strategic partnership on energy now forming between the U.S. and China is a promising start. Together, China and the U.S. are now the world's largest polluters. This suggests the need for a coordinated and vigorous energy policy. But an alliance between the U.S. and China is not enough. By 2050 the world's population is expected to reach nine billion, adding the equivalent of two Chinas to our current population. Nine billion people will need food, water, and other resources on a planet where human consumption is already negatively impacting the environment. The strategic planning necessary to move the global economy beyond fossil fuels remains elusive and will take leadership at the highest levels. As countries begin to search for ways to cooperate on comprehensive solutions to climate change, much more needs to be done to improve the performance of Green ICTs and reduce environmental pollution.

Conclusion

Managing sustainable economic growth and reducing environmental pollution are strong reasons to take the idea of a green economy seriously. Beyond economic nationalism, it is clear that the green economy necessitates a high degree of international cooperation. The rise of China and other emerging economies is a tectonic shift in the global order. Rising demand for energy in developing countries will be insurmountable without innovation in the global energy system. Beyond commercial markets, however, there is the need for coordination in policymaking to underwrite green innovation. Strategic public policy around green technologies is vital to reversing the negative impact of industrial production. Technology is no silver bullet, but ICTs are very likely the underlying foundations for harnessing an expanding green economy.

NOTES

1. The DOE has established the DOE Data Center Energy Efficiency Program.
2. Established in 1992 by the EPA, ENERGY STAR is the U.S. standard for energy efficient electronic equipment. Originally developed for computer equipment, it now includes other electronic equipment such as heating and cooling systems, office equipment, home electronics, etc.
3. DER technologies consist primarily of energy generation and storage systems such as solar, wind, fuel cells, and microturbines placed at or near the point of use.
4. In contrast, the U.S. share of total world energy consumption was 21 percent in 2006 but will fall to 17 percent in 2030.
5. During which the delegation proposed a list of 10 principles, some of which were incorporated into the UN's declaration later. Two environmental incidents related to water reservoirs (one in Dalian and one in Beijing) sparked the state's initial environmental consciousness and the development of a series of instruments including establish formal institutions, establishing environmental laws and organizations, and organizing large-scale environmental programs throughout the 70s and 80s (Economy, 2004).
6. In response to this growing problem, Tencent, one of China's largest Internet companies, has developed a next generation Green Data Center to manage its business services (Bie, 2008).

REFERENCES

Bauwens, M. (2009). Setting the broader context for P2P infrastructures: The long waves and the new social contract. Retrieved from: http://www.re-public.gr/en/?p=190

Bhanoo, S. (2010, Febuary 1). China's smart grid investments growing. *New York Times*. Retrieved from http://greeninc.blogs.nytimes.com/2010/02/01/chinas-smart-grid-investments-growing/

Bie, H. (2008). Tencent's "green" attempt. Retrieved from http://www.xinhuanet.com/topbrands/intel/greenit/article5.htm

Bradsher, K. (2010, January 30). China leading global race to make clean energy. *New York Times*. Retrieved from http://www.nytimes.com/2010/01/31/business/energy-environment/31renew.html?sq=us%20and%20china%20green%20economy&st=cse&adxnnl=1&scp=10&adxnnlx=1266095306-GyOow8G73xfxK91RYD31JA

Brodkin, J. (2009). Data center budgets rising despite economy, survey finds. *Network World*. 13 March, available at: www.networkworld.com/news/2009/031309-data-center-budgets-rising.html

Castro, D. (2009). Learning from the Korean green IT strategy. *The Information Technology and Innovation Foundation*. Retrieved from http://www.greenercomputing.com/research/report/2009/08/11/learning-korean-green-it-strategy.

China Beijing Environment Exchange (CBEEX) (2010). *CBEEX to launch China carbon neutral alliance*. Retrieved from http://www.cbex.com.cn/article//xxpd/bjsdt/201001/20100100015235.shtml

CCID (2009, January 19). *China's manufacturing industry development annual report*. Retrieved from http://www.ccidreport.com/report/content/6/200901/105586.html

Chaize, I. (2008). *Sharp increase in the amount of data in the 'digital universe.'* Security Watch, 13 March available at: www.securitywatch.co.uk/2008/03/13/sharp-increase-in-the-amount- of-data-in-the-digital-universe/.

CloudEx (2009). *China's internet in cloud computing era now.* Retrieved from http://www.cloudex.cn/index.html

CSIA (2002, July 24). *Guideline for vitalizing the software industry 2002–2005.* Retrieved from http://www.csia.org.cn/info/government/policy_statedepartment200247.htm

Department of Energy (DOE) (2009). U.S.-China clean energy announcements. *U.S. Department of Energy*, Retrieved from http://www.energy.gov/news2009/8292.htm

Economy, E. C. (2004). *The river runs black: The environmental challenge to China's future.* Ithaca, NY: Cornell University Press.

Energy Information Administration (2009). International energy outlook 2009. Retrieved from http://www.eia.doe.gov/oiaf/ieo/world.html

Gridwise (2009). "Smart grid." Retrieved Dec 19, 2009 from http://www.gridwise.org/

Gross, D. (2007). The U.S. is losing market share, so what? *New York Times*, January 28, BU5.

Hasson, J. (2009). *Data centers: Growth or stagnation?* FierceCIO, 15 March, available at: www.fiercecio.com/story/data-centers-growth-or-stagnation/2009–03–15

Hilty, L.M., Seifert, E.K., & Treibert R. (2005). *Information systems for sustainable development.* Hershey, PA: Idea Group Publishing.

Hu, J (2007, October 15). *Hu Jintao's Report on 17th National CPC Congress.* Retrieved from http://news.xinhuanet.com/newscenter/2007–10/24/content_6938568.htm

Jia, H. (2004, November 24). *China is second biggest greenhouse gas emitter.* Retrieved from http://www.scidev.net/en/news/china-is-second-biggest-greenhouse-gas-emitter.html

Jones, V. (2008). *The green-collar economy: How one solution can fix our two biggest problems.* New York: HarperOne.

Lu, X (2002). *Chinese informatization.* Beijing, China: Electronics Industry Publishing House.

Mandelson, P. (2010, Febuary 11). We want china to lead. *New York Times.* Retrieved from http://www.nytimes.com/2010/02/12/opinion/12ihtedmandelson.html?scp=1&sq=climate%20change%20and%20china&st=cse

Ministry of Environmental Protection (MEP) (July 25, 2007) *Environment Monitoring and Administration Methods.* Retrieved from http://www.gov.cn/ziliao/flfg/2007–08/07/content_708389.htm

Ministry of Environment, Republic of Korea, "Act on the Promotion of the Purchase of Environment-friendly Products," (enacted December 31, 2004) <www.koeco.or.kr/eng/download/green_procurement_law.doc> (accessed August 5, 2009).

National Development and Reform Commission (NDRC) (March 16, 2006). *The outline of the eleventh five-year plan: For national economic and social development of the people's republic of China.* Retrieved from http://ghs.ndrc.gov.cn/ghjd/115gyxj/001a.htm

People's Daily Online (PDO) (2009). People's Daily Online Investment in environmental protection to reach 3.1 trln yuan in the 12th five-year plan. People.com.cn. Retrieved from: http://english.people.com.cn/90001/90778/90862/6823402.html [Accessed 11 December 2009].

Rautenstrauch, C., and Patig, S. (2001). *Environmental information systems in industry and public administration.* Hershey, PA: IGI Global.

State Council of China (SCC) (2006). National guideline on medium- and long-term program for science and technology development. Beijing, China: SCC.

State Council of China (SCC) (2008). *National economy and social development informatization eleventh five year plan.* Beijing, China: SCC.

Shen, K., & Huang, G. (2009). *Comparative analysis of China's and India's information industry*. Retrieved from http://www.itxinwen.com/View/new/html/2009-03/2009-03-19-387521.html

Victor, D., & Yueh, L. (2010). The new energy order: Managing insecurities in the twenty-first century. *Foreign Affairs, 89,* 61–73.

Von Hippel, E. (2005). *Democratizing innovation*. Cambridge, MA: The MIT Press.

The World Bank Group (WBG) (2001). *China: Air, land, and water: Environmental priorities for a new millennium*. Washington, DC: World Bank Publications.

Worthington, T. (2009). *Green technology strategies: Using computers and telecommunications to reduce carbon emissions*. Belconnen, Australia: Tomw Communications.

Yong, Z. (September 23, 2009). *Construction of green data center is heating up in China*. Retrieved from http://it.sohu.com/20090923/n266936627.shtml

Yousif, M. (2010). *Towards Green ICT*. Retrieved from http://ercim-news.ercim.eu/en79/keynote/664-keynote-towards-green-ict

Afterword

SHOW US YOUR MESS!

An after-chat by:
Jean Burgess & Marcus Foth
Creative Industries Faculty
Queensland University of Technology
Brisbane, Australia

je.burgess@qut.edu.au
m.foth@qut.edu.au

MF: Do we have a structure for this?

JB: Yes...sort of...

MF: First, the problem with internet research as it is.

JB: I personally don't necessarily think we have to start with problems, but go for it.

MF: Then: the way out. Then: the book as a great example of that. Then: talk about the new "Internet studies" research with regard to the areas each of us works in.

JB: Well, I think the book hints at the problems as well as the solutions or, maybe more positively, hints at a question: what does internet research need?

MF: I never really questioned the raison d'être of internet studies, but maybe this is a good opportunity to do so.

JB: I have an answer in mind of course…it needs multidisciplinarity, and collaboration.

MF: The book's title, "New Intersections in Internet Research," hints at that, too. So why is that?

JB: Well, that is, if it is problem-based research, multi- or inter-disciplinary approaches are needed. Whereas if you are focused more on how your own research sits within relatively coherent disciplinary structures, then what internet research needs is whatever sociology/law/economics/computer science says "the discipline" needs next. But if you are trying to investigate a specific and more applied problem—such as sustainability, or piracy, or democracy online, then more often than not, no one discipline has all the tools and answers required to attack that problem.

MF: Hahahaha, but some of them think they do! ;-)—Right, so the big research questions are broad and encompass multiple domains of knowledge.

JB: But then how do you attack those questions? In contexts where the research is explicitly problem-based it often needs to be collaborative. I don't think that, for example, the extent of the collaboration we do here at QUT is just because we find it fun (although it is!)—for better or worse it's because of the more applied focus that structures our institutional context.

MF: I take a lot of this for granted by now until I visit other places and realise that our approach is actually quite progressive in comparison. I was never able to tick the box for "Pure Basic Research" to describe what I do, that is, "Experimental and theoretical work undertaken to acquire new knowledge without looking for long-term benefits other than the advancement of knowledge." It's always been applied, strategic, or experimental, but I have never been 'pure,' Jean! Oh no!

JB: Don't fret. I'm not pure either…Then of course there is the other side of that debate: that you need disciplinary depth to do quality and innovative research as well.

MF: So the breadth is about sustaining informed and mutually respected dialogue, whereas the depth comes from having a grounding in a particular field.

JB: Which also raises the question of balancing that disciplinary expertise with cross-disciplinary team-based work, particularly in doctoral research training. Your own field of "urban informatics" is a good example of an initiative that tries to do what we're talking about. If the "object" of research overall is people + place + technology . . .

MF: Well, yes, urban informatics is a creature of that process. The term is a convenient label, but didn't really exist with the current level of prominence back in 2006. On urbaninformatics.net, we say, "Our team comprises and collaborates with architects with degrees in media studies, software engineers with expertise in urban sociology, human-computer interaction designers with a grounding in cultural studies, and urban planners with an interest in digital media and social networking." I know Howard Rheingold has a nice explanation of "transdisciplinarity" somewhere, let me grab it quickly . . .

JB: Yeah, transdisciplinarity is probably the more appropriate term for some of these new research areas.

MF: According to Rheingold, "transdisciplinarity goes beyond bringing together researchers from different disciplines to work in multidisciplinary teams. It means educating researchers who can speak languages of multiple disciplines—biologists who have understanding of mathematics, mathematicians who understand biology."

[Quoted in a report by the Institute for the Future (IFTF) http://www.iftf.org/files/deliverables/SR1011_S&T_Map_2005–2055.pdf]

JB: While "interdisciplinarity" and "multidisciplinarity" are not the same thing, they both imply the continuation of disciplinary coherence, although brought into new dialogues. Whereas transdisciplinarity is about the development of the capacity to translate meaningfully across those boundaries, and also the development of literacies in disciplines other than your own—even outside of the social sciences in some cases.

MF: What say you? There is rigour outside the social sciences?? Just kidding…I think more and more colleagues not only lobby for this, they walk the walk. They publish not just in journals of their own home discipline, but branch out. The challenge is, how do you get accepted by those "others"?

JB: Well, this book is a good demonstration of one important shift that allows that co-authoring and collaboration across disciplines.

MF: Yeah, I love the fact that all chapters are co-authored. And that is a real achievement, since in a lot of places there is still widespread mutual disrespect.

JB: This is an example of fairly deep collaboration, where two or three authors work together on a conceptual or empirical problem and generate a coherent argument coming from different perspectives. And the interesting thing is that you will always end up with different answers and new problems that way.

MF: The disciplinary territory that the book covers is vast. It's a passport across linguistics, education, health, journalism, marketing, political science, technology, law, economics, the list goes on.

JB: Don't forget gender and cultural studies.

MF: What are they? ;-) But I think what's remarkable about the book's assemblage of research is that it does not only break down disciplinary silos and sits squarely across them, it also does away with many conventional dichotomies that have hampered progress and innovation, such as the chapters by Freeman and Cook and Monroy-Hernández, Dezuanni, and Kuikkaniemi that move beyond the simplistic split between online and offline.

JB: Yeah I agree, the title "Nexus" is spot on. It works on so many levels. Both Geneve & Ganito, and Nam and Camerini contribute different data pools from their respec-

AFTERWORD | 259

	tive case studies to their joint analysis and discussion. Many examine their topic from multiple theoretical perspectives. Breindl and Gustafsson's chapter is a great example.
MF:	There are also conNexions between different languages as demonstrated by Petzold & Liao. And, of course, geography: Australia and USA (Morieson & Usher), China and Taiwan (Houghton & Chang) ...
JB:	So the book presents not just a nexus of disciplines, but also of data sources, theories, dichotomies, languages, and places. Nice! But would you want to see us move to a totally post-disciplinary world, though?
MF:	Do they still have green tea ice cream in that world?
JB:	Because one of the functions of the disciplines is to develop communities of practice that agree on what a coherent relationship between theory, problem, and method is.
MF:	It needs to be more like a matrix organisation, more like a swarm I suppose. Or more problem-driven communities of practice that multiple disciplines contribute to. Take the sustainability issue for instance that Araya, Shang, and Liu discussed in their chapter.
JB:	Also, there should be more LOLcats.
MF:	Very true.

http://cheezburger.com/View.aspx?aid=3277126144

MF: But seriously, I think Internet researchers need at least some transdisciplinary education and research training. Such as supervisors from different fields.

JB: There's another reason we're even having this conversation I think. And that is that the 'internet' is so ubiquitous and embedded in so many more areas of social, cultural and economic life now, that many more fields of research are required to engage with it—urban planning, health, traditional humanities . . .

JB: We are starting to see more comparative and collaborative work across national boundaries, and that's really important as well—again something the chapters in the book demonstrate really well.

MF: Actually, it would be interesting to see how the different chapters came about…was there much fighting and/or debate and/or mutual explanation of discipline specific definitions going on when the authors wrote the chapters? I'd like to see a chapter that's the making of Nexus.

JB: "View source" scholarship! That's what I like about us adding our two cents in chat form—just for once it doesn't hide the messy process of bringing perspectives together to address a common issue.

MF: And could provide insights into how Academia could operate in a more transparent manner in the future.

JB: Ha! I think that's important with methodology as well—often we try to tidy up that mess before reporting our results. This (quite wonderful) book came about through the fairly random throwing-together of scholars from all over the globe; who moved beyond that original moment to set themselves the task of producing a book collaboratively. It would be wrong to try to impose a preexisting overarching logic on it, and actually this is a model of how knowledge creation works—through messy association, affinities, and networks.

MF: Yes, the process is messy! Show us your mess! There's a title for the Afterword.

Contributors

DANIEL ARAYA is a Ph.D. candidate in Educational Policy Studies at the University of Illinois (Urbana-Champaign). The focus of his research is the convergent impact of emerging technologies and cultural globalization on learning and innovation.

BRIDGET BLODGETT is currently a Ph.D. Candidate at the College of Information Sciences and Technology at the Pennsylvania State University. Her research involves the use of technology in virtual worlds and the social impacts of virtual worlds on offline life.

YANA BREINDL is a Ph.D candidate in Information and Communication Sciences at the Université Libre de Bruxelles (ULB) in Belgium. Her academic work deals with digital activism that aims at influencing European decision-making.

LUCA CAMERINI is a Ph.D candidate at the University of Lugano. He is currently involved in research activity with the Institute of Communication and Health, where he studies the impact of eHealth interactive applications on patients' affected by chronic conditions.

LUCY CRADDUCK is currently a doctoral candidate at the Faculty of Law, QUT and is undertaking a Doctorate of Juridical Science investigating the legal challenges facing the implementation of the proposed National Broadband Network in Australia. Lucy is a practicing lawyer in private practice and has worked in an in-house capacity with various Queensland Government departments.

YAO-CHUNG CHANG is a doctoral candidate at the ARC (Australian Research Council) Centre of Excellence in Policing and Security, Regulatory Institutions Network, at the Australian National University. His research is focused on the legal issues of information security, cybercrime, data protection, privacy, freedom of information, data retention, and telecommunication.

ERIC COOK is a doctoral candidate at the School of Information at the University of Michigan. Eric's research interests center on emergent social and creative behaviors afforded by technology placed in amateur, informal, and everyday contexts.

DR. MICHAEL DEZUANNI is a media literacy educator who lectures at Queensland University of Technology in the School of Cultural and Language Studies in Education. He recently completed a PhD in Education at QUT which focused on the performance of identities in media and technology education classrooms and the implications for participation with popular new media forms such as video games.

CARLA GANITO is a doctoral student in the Human Sciences Faculty at Portuguese Catholic University, where she also lectures on Digital Communication and Marketing. Her thesis addresses questions of gender and technology, focusing on the gendered use and representation of the mobile phone.

CRISTINA GARDUÑO FREEMAN is a PhD Candidate in the School of Architecture at the University of Technology Sydney. She is researching the way people engage with architecture through social media and online spaces, using the Sydney Opera House as a case study.

ANITZA GENEVE is a scholarship doctoral candidate in the Faculty of Science and Technology at Queensland University of Technology. Her Ph.D research topic explores the influences on women's participation in the multimedia and games production sectors of the Australian Digital Content Industry.

NILS GUSTAFSSON is a doctoral student in the Department of Political Science at Lund University, Sweden. His research focuses on the relationship between social media and political participation.

JULIAN HOPKINS is a scholarship doctoral student at Monash University, Sunway campus. He has degrees from Glasgow University, and from SOAS, University of London. His doctoral research involves participant observation with bloggers both on and offline, and the emerging phenomenon of the monetisation of blogging in Malaysia. You can visit his blog at www.julianhopkins.net.

TESSA J. HOUGHTON is a scholarship doctoral candidate in Media and Communication at the University of Canterbury, New Zealand. Her research investigates hacktivism through a neo-Habermasian model of the public sphere, following the radical or agonistic theoretical tradition.

KERK F. KEE (Ph.D. University of Texas at Austin, 2010) is Assistant Professor of Communication Studies at Chapman University in Orange County, California. His teaching and research focus on issues pertaining to innovation adoption, distributed collaboration, and virtual organizing in various contexts. His homepage is www.ekerk.com.

KAI KUIKKANIEMI is a doctoral student at the University of Arts and Design Helsinki. His research interests are experimental gaming, future of media business, and enactive media.

HAN-TENG LIAO is a doctoral student at the University of Oxford. His research aims to reconsider the role of keywords (sociolinguistics) and hyperlinks (webometrics) in shaping groups (governance or dynamic order) as bearers of ideas (political communication).

JINGFANG LIU is a PhD candidate at the Annenberg School for Communication and Journalism, University of Southern California. Her research interest is Information and Communication Technology and social change at the organizational, societal, and global levels.

ANDRÉS MONROY-HERNÁNDEZ is a PhD candidate at the MIT Media Lab doing research in social computing. As part of his work, he created the Scratch website, a large online community where young people share their own video games and animations.

LUCY MORIESON is a PhD candidate at RMIT University's School of Media and Communication in Melbourne, Australia. Her research is concerned with the cultural form and development of digital news, with a particular focus on relations between the newspaper and new forms of digital news.

YUJUNG NAM is a Ph.D. student at the University of Southern California's Annenberg School for Communication and Journalism. The focus of her research is the impact of a transforming media environment on the health communication domain and human-computer interaction.

RAMI OLWAN is an Australian Government PhD scholarship candidate at Queensland University of Technology (QUT). His current research focuses on the relationship between intellectual property, particularly copyright and development. He has published various articles in international journals on intellectual property and e-commerce laws. .

THOMAS PETZOLD is a doctoral candidate with the Centre of Excellence for Creative Industries and Innovation at Queensland University of Technology, Brisbane, Australia. His research focuses on the institutional and identity dynamics in the fields of language, governance, and digital culture, and looks in particular at the uses of digital multilingualism.

JIN SHANG (Ph.D. University of Leicester, 2010) recently completed a PhD in Mass Communications at the University of Leicester and is currently a research programme manager in the Department of Media and Communication. His main research interest focuses on investigating how ICTs and digital networks play a revolutionary role in socio-economic transformation, including related urban spatial restructuring, in an international context (digital economy & digital city).

NEAL THOMAS is a PhD candidate in the Art History and Communications Studies department at McGill University. His research is focused on the philosophical underpinnings of knowledge representation online.

NIKKI USHER is a PhD student at the University of Southern California's Annenberg School for Communication. She has been a Wallace Annenberg Fellow for the Center for Journalism and Democracy and a Knight Digital Media Center research assistant. Her work focuses on how traditional journalism is reshaping in a digital age.